GENETICS - RESEARCH AND ISSUES

MICRORNA AND NON-CODING RNA

TECHNOLOGY, DEVELOPMENTS AND APPLICATIONS

GENETICS - RESEARCH AND ISSUES

Additional books in this series can be found on Nova's website under the Series tab.

Additional e-books in this series can be found on Nova's website under the e-book tab.

GENETICS - RESEARCH AND ISSUES

MICRORNA AND NON-CODING RNA

TECHNOLOGY, DEVELOPMENTS AND APPLICATIONS

JAMES C. JOHNSON
EDITOR

New York

For permission to use material from this book please contact us:
Telephone 631-231-7269; Fax 631-231-8175
Web Site: http://www.novapublishers.com

Library of Congress Cataloging-in-Publication Data

Library of Congress Control Number: 2013934504

ISBN: 978-1-62618-442-8

Published by Nova Science Publishers, Inc. † New York

Contents

Preface

This book presents current research in the study of new technologies, developments and applications of microRNA and non-coding RNA. Topics include the role of microRNA and short interfering RNAs in plants; microRNAs, major affective disorders and suicidal behavior; microRNA in endoparasites; the functional roles of non-coding RNAs in glioma and their clinical implications; identifying miRNA function in innate immunity; and how multiple IsomiRs and diversity of miRNA sequences unveil evolutionary roles and functional relationships across animals.

Chapter 1 - In plants, two main categories of small RNAs (sRNAs) are distinguished based on their biogenesis and function: microRNAs (miRNAs) and short interfering RNAs (siRNAs). MiRNAs control the expression of cognate target genes by binding to reverse complementary sequences, resulting in cleavage or translational inhibition of the target RNAs. To date, nearly one thousand miRNA genes have been identified from various plant species as shown in miRBase and published reports, and this number is ever increasing. siRNAs have similar structure, function and biogenesis as miRNAs but are derived from long double-stranded RNAs and can often direct DNA methylation at target sequences. Plant miRNAs play important roles in maintenance of genome integrity and organ development such as leaf morphogenesis, floral organ identity and vascular development. The plant miRNAs also function in feedback regulation in the sRNA pathway and in directing biogenesis of some siRNAs. In this review, characteristics and functions, as well as signaling of miRNAs and siRNAs in plant have been summarized, with emphasis on their roles in plant developmental regulation of plants. Further studies on miRNAs and siRNAs are needed as they will not only provide us with important clues to understanding regulation mechanisms among miRNAs, siRNAs and their target genes, but also push forward future developments in this exiting field.

Chapter 2 - Major affective disorders are common and disabling conditions associated with significant psychosocial impairment and suicidal risk in the general population. At least 3–4% of all depressive individuals die by suicide. At a molecular level, affective disorders and suicidal behaviour have been recently associated with abnormalities in structural and synaptic plasticity. A recent hypothesis suggested that small non-coding RNAs (ncRNAs), in particular microRNAs (miRNAs), play a critical role in the translational regulation at the synapse. Here the authors summarized the current literature on miRNAs putative subcellular localization and sites of action in mature neurons analyzing their role in neurogenesis, synaptic plasticity, pathological stress changes, major affective disorders and suicidal behaviour. miRNAs have played a fundamental role in the evolution of brain functions. The perturbation of some intracellular mechanisms as well as impaired assembly, localization, and translational regulation of specific RNA binding proteins may affect learning and memory, presumably contributing to the pathogenesis of major affective disorders and perhaps suicidal behaviour. Also, miRNA dys-regulation has also been associated with several neuropsychiatric diseases. However, further evidence are needed in order to directly clarify the role of miRNAs in major affective disorders and suicidal behaviour [Abstract words (189); keywords (4): microRNAs; synaptic plasticity; major affective disorders; suicidal behaviour].

Chapter 3 - MicroRNAs (miRNAs) are short, non-coding RNAs that down-regulate gene expression post-transcriptionally by binding to the mRNA of their target genes. The importance of miRNAs in key biological processes such as development, cell proliferation, cell differentiation and metabolism has been widely documented since their discovery. To date, more than 25,000 mature miRNAs have been reported in 193 species of animals and plants (miRBase, release 19, http://www.mirbase.org/index.shtml). Parasites face different environments during their complex life cycles and their ability to respond to environmental and developmental signals may require miRNA-mediated networking and fine-tuning of gene expression. Some protozoan and helminth parasites have shown peculiarities in their small-RNA machinery. For example, the small RNA-generating machinery of the apicomplexan parasite *Toxoplasma gondii* is phylogenetically and functionally related to that of plants and fungi. Among protozoan parasites, miRNAs have been reported in *Giardia lamblia*, *Trichomonas vaginalis*, *Trichomonas foetus*, *Pentatrichomonas hominis, Trypanosoma brucei*, *Entameoba histolytica*, *Neospora caninum* and *Toxoplasma gondii*. Most of the protozoan miRNAs do not share significant homology to any of the known miRNAs of plants and metazoans. Some interesting roles such as silencing of variant surface proteins have been proposed for a number of miRNAs in *G. lamblia*. However, in other unicellular parasites such as *Trypanosoma cruzi* and *Plasmodium falciparum* no miRNAs or small RNA machinery components have been identified. Among helminth parasites, miRNA identification and expression profile analyses have been performed by a combination of bioinformatic and experimental approaches. So far, miRNAs have been reported in the nematodes *Angiostrongylus cantonensis*, *Brugia pahangi*, *Brugia malayi*, *Haemonchus contortus*, *Ascaris suum* and *Trichinella spiralis;* the trematodes *Fasciola hepatica*, *Fasciola gigantica*, *Schistosoma mansoni*, *Schistosoma japonicum*, *Clonorchis sinensis and Orientobilharzia turkestanicum* and the cestodes *Echinococcus granulosus, Echinococcus multilocularis* and *Taenia saginata*. Conserved as well as novel miRNAs have been reported, some of them are abundantly expressed and display developmental regulation, suggesting important roles in several helminth parasite species. miRNA targets have been bioinformatically predicted in parasitic protozoa and helminths but only a small number of protozoan miRNAs has been experimentally validated. Unraveling miRNA roles in parasite biology and host interaction remains a challenge. For this end, bioinformatic strategies adapted to each phylum together with experimental strategies such as miRNA silencing and/or overexpression should be set up. The knowledge of miRNA roles in parasite development and survival in their hosts will in turn provide new control and intervention strategies against the worldwide distributed and mostly neglected diseases they produce.

Chapter 4 - Glioma is the commonest form of primary brain tumors with poor clinical outcome. MicroRNAs (miRNAs) are endogenous non-coding RNAs (ncRNAs) of approximately 22 nucleotides (nt) that regulate the expression of target messenger RNAs (mRNAs) at the post-transcriptional level. Increasing evidence shows that miRNAs play key roles in the biological behavior of glioma. They are involved in a wide variety of gliomagenic processes including cell proliferation, invasion and apoptosis by acting as tumor suppressors or oncogenes. MiRNAs may potentially be exploited as biomarkers for diagnosis, prognostication, and even therapeutic targets. Recently, the characterization of long non-coding RNAs (lncRNAs) has received considerable attention. Like their smaller non-coding miRNA counterparts, lncRNAs represent a novel class of gene regulators in cancer. Early evidence demonstrates that the differential expression of specific lncRNAs may also serve as novel biomarkers for diagnosis and therapies. In this chapter, the authors summarize the recent advances in the investigations of miRNAs and lncRNAs in glioma, and discuss their potentials for clinical use.

Chapter 5 - miRNA regulates immunity by modulating the expression of key genes involved in leukocyte development and function. Furthermore, immunity includes innate and adaptive systems comprising myeloid-derived cells and lymphocytes which mediate nonspecific and pathogen-specific responses. Innate immunity encompasses the generation of myeloid-derived cells and their ability to mount a rapid and appropriate response to pathogen-associated danger. Here the authros highlight common components of research investigating the role of miRNA in innate immunity. The identification of miRNAs involved in immunity generally begins with the profiling of samples generated ex vivo or in vitro which include the cell types and immunological conditions of interest. From here a core set of methods and technologies are utilised. Various technologies exist for quantifying miRNA expression, including qRT-PCR and microarray platforms capable of profiling all known human miRNAs as listed by miRBASE. miRNA array expression data may be confirmed by qRT-PCR before in silico analysis of predicted miRNA targets, and these miRNA-target interactions validated by dual luciferase assays. Finally, the biological significance of any miRNA is dependent upon the degree of modulation it exerts combined with the importance of it target within the relevant system. A widely used approach for ascertaining this is to perform experiments in which the miRNA is constitutively expressed or knocked-down and altered function or experimental outcome tested. To date these methods have revealed several important roles for miRNA in the functioning of the immune system. As the technologies and techniques described here continue to develop so will our understanding of the function of miRNA within immunity and with our prospects of manipulating expression for therapeutic gain.

Chapter 6 - As crucial negative regulatory molecules, small non-coding RNAs, microRNAs (miRNAs), have been widely studied, especially for their functions and evolutions. Multiple miRNA variants, termed as isomiRs, are recently detected and identified from a given miRNA locus due to alterative and imprecise cleavage of Drosha and Dicer. Compared to the canonical miRNA sequence, these isomiRs may be found various 5' and/or 3' ends (various 3' ends are more popular), 3' additional non-template nucleotides, 5' and seed modification events. Similarly, miRNA sequences in different animal species also show diversity as well as species specificity. The miRNA sequences in different animal species may be involved in different length distributions, 5' and/or 3' ends, transition/transversion and insertion/deletion. From a given miRNA locus, internal multiple isomiRs and various miRNA sequences across species are quite similar and have close relationships. Indeed, they also implicate potential functional and evolutionary relationships. By applying high-throughput sequencing technique, multiple isomiRs from a given miRNA locus can be identified with different sequence counts, length distributions and various expression patterns, although isomiR repertoire and their expression patterns always are stable across different tissues, samples and even animal species. Most isomiRs were 3' isomiRs which had the same seed sequences, while 5' isomiRs were found to have new seed sequences via the phenomenon of "seed shifting". Many miRNAs were detected a set of isomiRs, in which 1-3 were abundant isomiRs characterized by a narrow length distribution (20-23 nt) due to dominant cleavage. Indeed, specific miRNA sequences have been identified in specific animal species through classical experimental validation. The annotated miRNA sequence is always derived from the most abundantly expressed isomiR, which is prone to be detected by traditional detection. Therefore, the authors speculate that the most dominant isomiR is not randomly selected, but is strictly regulated by complex regulatory network from selection and evolutionary pressures. Various miRNA sequences can unveil dynamic evolutionary processes across different animal species (cross-species) and across different miRNAs (within species), especially simultaneously involved in the diversity of pre-miRNA and pri-miRNA sequences. Furthermore, more miRNA* sequences, ever as passenger strands of miRNA sequences, are identified as active and functional miRNA sequences. These potential functional miRNA* sequences and potential more functional isomiRs (such as 5' isomiRs with novel seed sequences by changing 5' ends) will complicate miRNA-mRNA and miRNA-miRNA interaction, and enrich coding-non-coding RNA regulatory network.

In: MicroRNA and Non-Coding RNA
Editor: James C. Johnson

ISBN: 978-1-62618-442-8

Chapter 1

Role of MicroRNAs and Short Interfering RNAs in Plants

Chunhua Zhang*[1]*, Zhijun Shen*[1]*, Ruijuan Ma*[1]*,
***Nicholas Kibet Korir*[2] *and Mingliang Yu*[1*]**
[1]Institute of Horticulture, Jiangsu Academy of Agricultural Sciences, Nanjing, P. R. China
[2]Department of Agricultural Technology, Kenyatta University, Nairobi, Kenya

Abstract

In plants, two main categories of small RNAs (sRNAs) are distinguished based on their biogenesis and function: microRNAs (miRNAs) and short interfering RNAs (siRNAs). MiRNAs control the expression of cognate target genes by binding to reverse complementary sequences, resulting in cleavage or translational inhibition of the target RNAs. To date, nearly one thousand miRNA genes have been identified from various plant species as shown in miRBase and published reports, and this number is ever increasing. siRNAs have similar structure, function and biogenesis as miRNAs but are derived from long double-stranded RNAs and can often direct DNA methylation at target sequences. Plant miRNAs play important roles in maintenance of genome integrity and organ development such as leaf morphogenesis, floral organ identity and vascular development. The plant miRNAs also function in feedback regulation in the sRNA pathway and in directing biogenesis of some siRNAs. In this review, characteristics and functions, as well as signaling of miRNAs and siRNAs in plant have been summarized, with emphasis on their roles in plant developmental regulation of plants. Further studies on miRNAs and siRNAs are needed as they will not only provide us with important clues to understanding regulation mechanisms among miRNAs, siRNAs and their target genes, but also push forward future developments in this exiting field.

1. Introduction

Small non-coding RNAs, which consist of 20–24 nucleotides (nt), have been increasingly investigated as important regulators of protein coding gene expression. They were first reported in the nematode *Caenorhabditis elegans* (*C. elegans*) and are responsible for the phenomenon known as RNA interference (RNAi), co-suppression, gene silencing, or quelling (Lee et al., 1993; De Carvalho et al., 1992; Hannon, 2002; Napoli et al., 1990; Romano and Macino, 1992). These small RNAs (sRNAs) function by causing either transcriptional gene silencing (TGS) or posttranscriptional gene silencing (PTGS) (Baulcombe, 2004). The diversity and widespread existence of sRNAs and the fundamental importance of sRNA-mediated regulation has recently attracted numerous comprehensive studies in both animals and plants. SRNAs are involved in virtually all aspects of plant biology by interfering with mRNA translation, including development, responses to biotic and abiotic stresses, nutrient allocation and maintenance of genome integrity (Chitwood and Timmermans, 2010; Chuck et al., 2009; Sunkar et al., 2012).

MicroRNAs (miRNAs) and small interfering RNAs (or 'short interfering RNA', siRNAs) are two main categories of sRNAs in plants, and are distinguished by their different function and modes of biogenesis (Axtell et al., 2011; Chapman and Carrington, 2007; Melnyk et al., 2011; Voinnet, 2009). MiRNAs are about 22 nt long endogenous, non-translated RNAs, which regulate many biological processes in plants by mediating translational inhibition or cleavage of target transcripts (Kim et al., 2008). Although miRNAs were first discovered two decades ago (Lee et al., 1993; Wightman et al., 1993), they were not recognized until 2001 by Lee and other scientists who reported that *lin-4* and *let-7* antisense RNAs are temporal regulators that control the timing of developmental events by inhibiting translation of target mRNAs in *C. elega* (Lee and Ambros, 2001; Lau et al., 2001; Lagos-Quintana et al., 2001). Since then, miRNAs have attracted huge interest from scientists in the world (Bartel, 2004). The current miRNA database 19 (miRBase 19) (http://www.mirbase.org/ftp.shtml), released in August 2012, contains over 5940 mature miRNA sequences in over 67 plant species (http://www.mirbase.org/ftp.shtml, Kozomara and Griffiths-Jones, 2011). based on the miRBase 19 statistics, about 320 plant miRNA families have been identified so far, with more than half of these being found in the fully sequenced genomes of *Arabidopsis thaliana* (*A. thaliana*), *Populus trichocarpa* (*P. trichocarpa*), and rice (*Oryza sativa*). Many plant miRNAs are well conserved in multiple species, indicating that the miRNA-directed gene regulation process is very ancient (He et al., 2008). Several laboratory and computational approaches have shown that this class of sRNAs has diverse and important roles in plant growth and development (Bartel, 2004; Zhang et al., 2006). Moreover, a wealth of studies further established that hundreds of eukaryotic miRNAs seem to regulate a vast array of other biological functions including hormonal control, immune responses, and adaptation to a variety of biotic and abiotic stresses (Zhang et al., 2006; Sunkar et al., 2007; Voinnet, 2008; Pedersen and David, 2008). SiRNA can be derived from the genome (as shown in plants) or exogenously supplied in the form of a long duplex of nucleic acids. They are found in most eukaryotes, and are the most abundant sRNAs in plants (Mette et al., 2002; Llave et al., 2002). The siRNAs were first discovered in plants exhibiting transgene- or virus-mediated RNA silencing (Hamilton and Baulcombe, 1999) with endogenous siRNAs having been classified into at least three subclasses: repeat-associated siRNAs (ra-siRNAs), trans-acting siRNAs (ta-siRNAs), and siRNAs derived from natural antisense transcripts (nat-siRNAs) (Sunkar et al., 2007).

SiRNAs are closely related to miRNAs but they differ in terms of their origin, structure, associated effector protein and mode of action. There are key differences between miRNAs and siRNAs in origin and evolutionary conservation: (1) miRNAs derive from genomic loci distinct from other recognized genes, whereas siRNAs derive from mRNAs, transposons, viruses, or heterochromatic DNA. (2) miRNAs are processed from transcripts that can form local RNA hairpin precursor structures, whereas siRNAs are processed from long bimolecular RNA duplexes or extended hairpins. (3) a single miRNA molecule ultimately accumulates from one arm of each miRNA hairpin precursor molecule, whereas many different siRNAs accumulate from both strands of siRNA precursors. (4) miRNA sequences are nearly always conserved in related organisms, whereas siRNA sequences are rarely conserved in related organisms. These differences have been used to develop practical guidelines for distinguishing and annotating newly discovered miRNAs (Bartel and Bartel, 2003; Ambros et al., 2003). This review summarizes the nature, biogenesis and various functions of miRNAs and siRNAs on plant growth, development, and stress responses, as well as signaling in plant. Due to the multiple functions of miRNAs and siRNAs studied before, one can anticipate that every laboratory working on plant development, stress or virus resistance will now look for the possible involvement of miRNA and siRNAs in the processes, thus providing further insights into the understanding of sRNA-mediated regulation and its importance in plant biology.

2. MiRNA

2.1. MiRNA Biogenesis

MiRNA biogenesis is a multi-step process including transcription, processing, modification, and RISC loading (Figure 1). In plants, miRNAs are processed from single-stranded hairpin precursors ranging between 64-303 nt. Genes encoding miRNAs in plants are annotated as *MIR* genes. A nuclear-encoded *MIR* gene is transcribed by RNA polymerase II (RNA Pol II) into a capped and polyadenylated primary miRNAs (pri-miRNA) with a characteristic hairpin sturcture, which is processed into the stem-loop precursor, pre-miRNA, by a Dicer-Like1 (DCL1) protein (Chen, 2009; Park et al., 2002; Reinhart et al., 2002; Xie et al., 2005). This latter step also requires the cooperation of the Hyponastic Leaves1 (HYL1), Serrate (SE) and Cap-Binding proteins (CBP20 and CBP80). The pre-miRNA hairpin precursor is further processed by the dicer activity of the DCL1/HYL1 complex and SE into a 20 to 22 nt duplex of miRNA/miRNA* (Chen, 2009; Garcia, 2008; Kim et al., 2008). This duplex is composed of the mature miRNA strand and its complementary strand, commonly noted as miRNA*. The miRNA/miRNA* duplex needs to be protected from degradation by addition of methyl groups on the 3'ends. The miRNA/miRNA* is methylated on the 2'OH of the 3' terminal nucleotides by Hua Enhancer1 (HEN1) methyl transferase which has the enzymatic activity required for this step (Yu et al., 2005). Methylated miRNA duplex is exported from the nucleus to the cytoplasm through export factors, Hasty (HST1) (Kim, 2004; Park et al., 2005; Khraiwesh et al., 2012).

Within the cytoplasm the two strands of the miRNA duplex are separated by helicase and a single-stranded mature miRNA is loaded into an Argonaute (AGO) protein (generally AGO1) contained in miRNA RISC (miRISC), while the miRNA* strand is degraded.

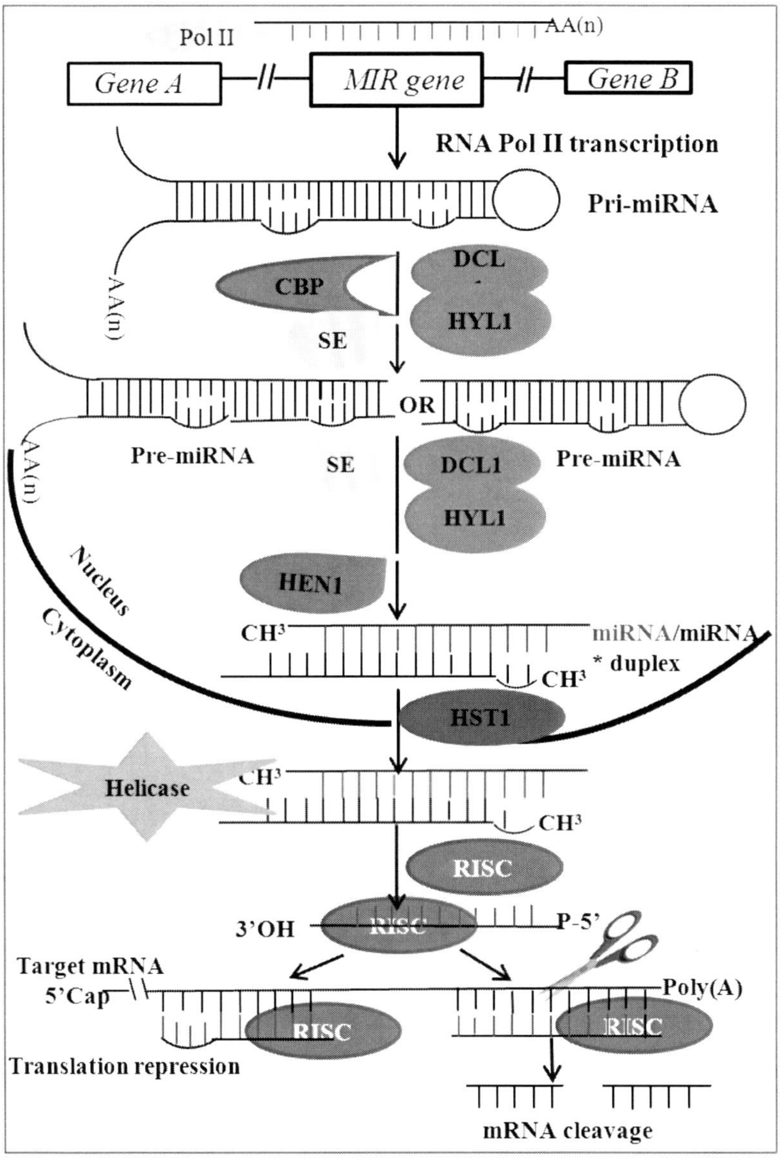

Figure 1. Biogenesis and function of miRNAs. *MIR* genes are initially transcribed by Pol II into a single-stranded RNA that folds back to form a hairpin structure (pri-miRNA). Splicing pri-miRNA and further processing to be pre-miRNA in nuclear dicing bodies involve the interactive functions of HYL1 and SE and of the cap-binding proteins (CBP) CBP20 and CBP80. The pre-miRNA hairpin precursor is further processed by the Dicer activity of the DCL1/HYL1 complex and SE into a 20 to 22 nt duplex of miRNA/miRNA* duplexes. These are then methylated by HEN1 and transported into the cytoplasm by HST1. The miRNA guide strand is selected, incorporated, and stabilized in dedicated AGO1 protein. MiRNA-guided AGO1-containing RISC directs mRNA cleavage or translation repression of the target mRNA.

The current model postulates that mature miRNAs are incorporated into the RNA-induced silencing complex (RISC), bind target mRNAs based on complementarity, and guide cleavage of mRNA targets with perfect or near perfect complementarity and translational repression of targets with lower complementarity (Bartel, 2004). The identity and activities of RISC components are incompletely understood, but in all eukaryotes examined, a member of the AGO protein family is a principal RISC component (Hammond et al., 2001). AGO proteins, containing amino-terminal PAZ (for PIWI/Argonaute/Zwille) domains and carboxy-terminal PIWI domains, are core components of the RNAi mechanisms (Parker et al., 2005). Rand et al. (2004) also reported that AGO proteins seem to be the sole proteins required for RISC-mediated activities. The AGO protein induces the sequence specific repression of the miRNA target genes, either by cleavage or translational repression. Baumberger et al. (2005) reported that AGO1 protein is an effector enzyme of RISC and is known to exhibit the "slicer" activity required for target cleavage. RISC cleaves the target mRNA strand between positions 10 and 11, counted from the 5' to 3' end of the guide RNA (Rand et al., 2004). Active RISCs, which contain only a single guide RNA strand, are capable of directing multiple rounds of site-specific target mRNA cleavage. The mode of translational repression depends on the degree of sequence complementarity between the miRNA guide strand and the target mRNA sequence. Nucleotides 2–8 (the 'seed' region) of a miRNA are almost always perfectly complementary to a mRNA if it is to target the mRNA for translational repression (Snead and Rossi, 2010). In this way, gene expression is regulated by miRNAs through prevention of successful translation in a target mRNA.

Several different mechanisms on how miRNAs achieve noncleavage translational repression have been proposed. MiRNA-mediated translational repression may be occurring at the translational initiation stage and the elongation stage, by facilitating posttranslational polypeptide degradation, and/or a spatial or temporal combination of these. Evidence for each of these potential mechanism has been reported.

2.2. MiRNA Functions

2.2.1. MiRNA in Response to Plant Development and Evolution

Leaf: It is now clear that many plant developmental processes are regulated by miRNAs (Mallory and Vaucheret, 2006). Palatnik et al. (2003) reported that miRNA-mediated control of leaf morphogenesis is conserved in plants with very different leaf forms. In both monocots and dicots, regulation by miR165/miR166 is required for organ axis specification, for vascular development and for meristem function (Juarez et al., 2004; Kidner and Martienssen, 2004; Mallory et al., 2004b; Emery et al., 2003). MiR166 can promote abaxial/ventral fate in developing primordia by restricting the expression of class III homeodomain leucine zipper (HD-ZIP III) transcription factors, which are necessary and sufficient to specify leaf polarity in maize (McConnell et al., 2001; Emery et al., 2003; Juarez et al., 2004). In *Landsberg erecta*, it was also found that spatially restricted miR165/miR166 directs leaf polarity through AGO1 (Kidner et al., 2004) (Table 1).

It was identified that miR396 has two loci (miR396a and miR396b) and targets six *Growth-Regulating Factor* (*GRF*) genes encoding putative transcription factors with roles in *Arabidopsis* leaf growth. To analyze the role of miR396 in the regulation of target genes and leaf growth, Liu et al. (2009) have generated transgenic *Arabidopsis* plants that constitutively overexpress miR396a or miR396b. These transgenic plants have narrow-leaf phenotypes due to reduction in cell number. Ectopic overexpression of miR396 represses expression of not only six *GRF* genes but also *GIF1* (*GRF-interacting transcription*), encoding a GRF-interacting transcription coactivator with a role in cell proliferation in leaf. In addition, transgenic MIR396-overexpressing plants have lower densities of stomata and are more tolerant to drought than wild-type plants. These results strongly support the belief that miR396 plays an important role in plant leaf growth and development, most likely by repressing *GRF* gene expression. Rodriguez et al. (2010) also observed that the balance between miR396 and the *GRFs* controls the final number of cells and cell proliferation in *Arabidopsis* leaves. MiR396 can regulate cell proliferation and the size of the meristem. Overexpression of miR396 in a mutant lacking *GIF1* severely compromises the shoot meristem.

MiRNAs regulate many genes, and the readout is complex. Of the 24 *A. thaliana TCP* genes, five contain a target site for miR319: *TCP2* (*At4g18390*), *TCP3* (*At1g53230*), *TCP4* (*At3g15030*), *TCP10* (*At2g31070*), and *TCP24* (*At1g30210*). The miR319 regulatory network has been implicated in different aspects of leaf development, from morphogenesis to leaf senescence and from cell proliferation to cell differentiation.

The *TCP4* mutant and *jaw-D* display a moderate late-flowering phenotype with an increase in leaf number to 22 compared to 15 in wild type (Palatnik et al., 2003; Schommer et al., 2008; Sarvepalli and Nath, 2011; Schommer et al., 2012)

Embryo (cotyledon): MiR164 can limit the expansion of the boundary size in meristem and control embryo patterning through mediating the degradation of *CUP-SHAPED COTYLEDON 1* (*CUC1*) and *CUC2* mRNAs. As for seedling phenotype, the wild-type seedlings have aligned cotyledons misaligned, while the transgenic lines showed partially fused or cup-shaped cotyledons and exhibited petiole fusion (Laufs et al., 2004). MiR164-directed regulation of target genes *CUC* mutant causes development abnormalities in embryonic organs with upright cotyledon petioles, thus indicating miR164 as a key regulatory component essential for normal embryo development (Mallory et al., 2004b). Palatnik et al. (2007) reported that patterning defects, including fused cotyledons, have also been observed in *soj* (for *suppressor of jaw-D*) plants, which have mutations in the *miR319* binding site of *TCP4*. Koyama et al. (2007) proposed that at least part of these patterning defects is caused by the repression of *CUC* genes by the *TCPs*.

Flower and developmental clocks: The analysis of cincinnata mutants in *Antirrhinum* showed that in addition to affecting leaf growth, the miRNA-regulated *TCP* gene, *CINCINNATA*, affects petal lobe development by controlling epidermal cell differentiation and growth (Crawford et al., 2004). *MiR319a* loss-of-function allele was isolated in *Arabidopsis* (Nag et al., 2009). The *miR319a* mutants exhibit defects in petal and stamen development, presenting narrower and shorter petals as well as impaired anther formation (Nag et al., 2009). The petals are greenish in the *jaw-D* mutant, suggesting that a decrease in *TCP* level also affects petal development (Palatnik et al., 2003; Weigel et al., 2000).

Table 1. Functions of miRNAs involved in plant development

Functions in	MiRNA	Target gene	Plant species	Reference
Leaf margin serration	miR164a	*NAC* (*CUC2*)	*Arabidopsis*	Nikovics et al., 2006
Leaf,floral,embryo development	miR164	*NAC* (*CUC1*)	*Arabidopsis*	Mallory et al., 2004b
Leaf development	miR165/166	*HD-ZIPIII* (*PHB,PHV*)	*Arabidopsis*	Mallory et al., 2004a
Leaf polarity	miR165/166	*HD-ZIPIII*	maize	Juarez et al., 2004
Leaf polarity	miR165/166	*HD-ZIPIII*	*Landsberg erecta*	Kidner et al., 2004;
Shoot meristem and leaf axial patterning	miR165/166	*HD-ZIPIII*	*Arabidopsis*	Grigg et al., 2005
Leaf senescence	miR319	*TCP*	*Arabidopsis*	Schommer et al., 2008
Leaf morphlogenesis	miR159a,b	*MYB33* and *MYB65*	*Arabidopsis*	Allen et al., 2007
Leaf growth (shape)	miR396	*GRF*	*Arabidopsis*	Liu et al., 2009
Cell proliferation and the size of the meristem of leaf	miR396	*GRF*	*Arabidopsis*	Rodriguez et al., 2010
Leaf morphogenesis	miR319	*TCP*	*Arabidopsis*	Palatnik et al., 2003; Schommer et al., 2012
Leaf growth and petal lobe development	miR319	*TCP*, *CINCINNATA*	*Antirrhinum*	Crawford et al., 2004
Axillary meristem for shoot or flower	miR164	*NAC* (*CUC1* and *CUC2*)	*Arabidopsis*	Raman et al., 2008
Petal and stamen development	miR319a	*TCP4*	*Arabidopsis*	Nag et al., 2009
Inflorescence stems	miR166	*HD-ZIP III*	*Arabidopsis*	Kim et al., 2005
Stamen development	miR159	*GAMYB* (*MYB33* and *MYB65*)	*Arabidopsis*	Millar and Gubler, 2005
Stamen development	miR159	*GAMYB*	rice	Kaneko et al., 2004
Stamen development	miR159	*GAMYB*	*Arabidopsis*	Achard et al., 2004; Schwab et al., 2005; Palatnik et al., 2007
Leaf growth and petal lobe development	miR319	*TCP*, *CINCINNATA*	*Antirrhinum*	Crawford et al., 2004
Floral patterning	miR172	*AP2*	*Nicotiana benthamiana*	Mlotshwa et al., 2006
Petal number	miR164c	*NAC* (*CUC1* and *CUC2*)	*Arabidopsis*	Baker et al., 2005
Boundary size control of meristems	miR164	*CUC1,CUC2*	*Arabidopsis*	Laufs et al., 2004; Mallory et al., 2004b

Table 1. (Continued)

Functions in	MiRNA	Target gene	Plant species	Reference
Ovule and anther development	miR167	*ARF6 and ARF8*	*Arabidopsis*	Wu et al., 2006
Seedling development	miR319	*TCP4*	*Arabidopsis*	Schommer et al., 2012
Embryo patterning	miR319	*TCP4*	*Arabidopsis*	Schommer et al., 2012
Cotyledons	miR164	*NAC* (*CUC1 and CUC2*)	*Arabidopsis*	Laufs et al., 2004
Core components of the clock	miR319	*TCP2, TCP3*	*Arabidopsis*	Giraud et al., 2010
Flowering time and floral organ identity	miR172	*AP2*	*Arabidopsis*	Aukerman and Sakai, 2003
Flowering time and leaf identity	miR172	*AP2*	maize	Lauter et al., 2005
Flowering time and anther development	miR159	*GAMYB* (*DELLA*)	*Arabidopsis*	Achard et al., 2004; Schwab et al., 2005
Root cap formation	miR160	*ARF10, ARF16*	*Arabidopsis*	Wang et al., 2005
Development of adventitious roots	miR160	*ARF17*	*Arabidopsis*	Mallory et al., 2005
Nodule and root development	miR166	*HD-ZIP III*	*M. truncatula*	Boualem et al., 2008
Differentiation of nodule cells	miR169	*CCAAT-binding family* (*MtHAP2-1*)	*M. truncatula*	Combier et al., 2006

ARF6 and *ARF8* transcripts are cleavage targets of the *miR167*, and overexpressing *miR167* mimicked *arf6 arf8* phenotypes. Wild-type pollen grew poorly in *arf6 arf8* gynoecia, correlating with *ARF6* and *ARF8* expression in style and transmitting tract. Mutations in the *miR167* target sites of *ARF6* or *ARF8* caused ectopic expression of these genes in domains of both ovules and anthers where *miR167* was normally present. As a result, ovule integuments had arrested growth, and anthers grew abnormally and failed to release pollen (Wu et al., 2006).

MiR164c can control petal number in a nonredundant manner by regulating the transcript accumulation of the transcription factors CUC1 and CUC2 (Baker et al., 2005), and also constrains the expansion of the boundary domain in *Arabidopsis* meristems by degrading *CUC1* and *CUC2* mRNAs with wild-type or control lines showing unfused sepals and expanded petals and stamens (Laufs et al., 2004). However, the transgenic lines (*2x35S::miR164A* and *2x35S::miR164B*) have fewer petals and the margins of two adjacent sepals are quite close thus hiding the petal insertion point. Similar sepal fusion and reductions in petal number were reported for flowers of the *cuc1* and *cuc2* double mutant (Aida et al., 1997). The formation of axillary meristem which develops into side shoots or flowers is reported to be regulated by miR164-mediated *CUC1* and *CUC2* in *Arabidopsis* (Raman et al., 2008). However, according to the observation on miR164-directed *CUC* gene by Mallory et al. (2004b), the mutants have flowers with six petals and two sepals compared with control plants which have four petals and four sepals,. MiR319a targeting of *TCP4* affects petal and stamen development with petals of *miR319a* mutants being narrow and short, while stamens exhibit defects in anther development (Nag et al., 2009).

MiR166-mediated *ATHB15* mRNA cleavage plays important roles in the regulation of vascular development of inflorescence stems. In a gain-of-function *MIR166a* mutant, the decreased transcript level of *ATHB15* was accompanied by a fasciated inflorescence stem with disrupted radial vascular patterning and expanded xylem tissue in *Arabidopsis* (Kim et al., 2005). More flowers were produced in mutants than in wild-type plants. In addition, the carpels of mutant are very short and sterile, with a slight downward curling leaf.

The regulation of the floral homeotic gene *APETALA2* (*AP2*) by miR172 is crucial for normal flower development and is likely to be conserved across plant species, such as in *Arabidopsis*, maize and *Nicotiana benthamiana*. Expression of *miR172*-resistant *Arabidopsis* AP2 in *N. benthamiana* (*35S::AP2m3* lines) induced floral patterning defects and the severity of the defects correlated with AP2 mRNA and protein levels. Mlotshwa et al. (2006) found that the phenotypes were characterized by reduced petal size and short-and-bulged carpels in at least 60% of the flowers of most *35S::AP2m3* lines. Some *35S::AP2m3* lines produced flowers with numerous stamens and carpels, followed by emergence of numerous petals in all flowers in late stages of development. All the flowers produced by these lines were severely phenotypic and completely sterile (Mlotshwa et al., 2006).

Similar to the role of certain miRNAs for developmental timing in *C. elegans*, some plant miRNAs also function as developmental clocks. Multiple examples show that miRNAs can efficiently control developmental timing in plants (Garcia, 2008). In *A. thaliana*, the targeting of *AP2-like* genes by miR172 is controlling the transition from vegetative to reproductive phase (Aukerman and Sakai, 2003; Chen, 2004). The progressive accumulation of miR172 during vegetative development induces a gradual decrease in the abundance of its targets. When the target reaches a critical threshold, it promotes flowering.

Lauter et al. (2005) reported that the expression level of AP2-like gene *glossy15* determines the length of the juvenile phase in maize. The increase in miR172 accumulation progressively represses *glossy15* during juvenile development. When a *glossy15* threshold is reached, the apparition of adult traits is promoted.

Flowering time is also controlled by the miR156 family. Overexpression of miR156 leads to the repression of transcription factors of the SBP and SPL families, extends juvenile development and delays flowering time (Schwab et al., 2005; Wu and Poethig, 2006). Yamaguchi et al. (2009) reported that all three genes *LEAFY* (*LFY*), *FRUITFULL* (*FUL*) and *APETALA1* (*AP1*) are directly activated by the miRNA targeted transcription factor SQUAMOSA PROMOTER BINDING PROTEIN-LIKE 3 (SPL3). These findings suggest that SPL3 acts together with other miRNA-regulated SPL transcription factors to control the timing of flower formation (Yamaguchi et al., 2009). Furthermore, Ori et al. (2007) demonstrated that miR319 plays a role in the timing of morphogenesis versus differentiation of leaf margins.

MiR159 plays a role in the regulation of short-day photoperiod flowering time and of anther development by directing the cleavage of mRNA encoding GAMYB-related proteins (Achard et al., 2004). Increased level of miR159 cause a reduction in *LEAFY* transcript level, delay flowering in short-day photoperiods, and upset anther development in transgenic *Arabidopsis*.

Root: Several studies report that miRNAs are involved in the development of root and nodule (Boualem et al., 2008; Combier et al., 2006). In the model legume *M. truncatula*, miRNAs post-transcriptionally regulate the *HD-ZIP III* genes, a new family of transcription factors associated with nodule development. These target genes are spatially co-expressed with miR166 in vascular bundles, and in apical regions of roots and nodules. MiR166 overexpression reduces the number of symbiotic nodules and lateral roots, and induces ectopic development of vascular bundles in the transgenic roots (Boualem et al., 2008). Using *M. truncatula* Combier and coworkers (2006) demonstrated that *MtHAP2-1*, a new transcription factor of the CCAAT-binding family, is regulated by miRNA169, whose overexpression leads to the same nodule developmental block as *MtHAP2-1* RNAi constructs. The complementary expression pattern of miR169 and *MtHAP2-1* and the phenotype of miR169-resistant *MtHAP2-1* nodules strongly suggest that the miR169-mediated restriction of *MtHAP2-1* expression to the nodule meristematic zone is essential for the differentiation of nodule cells.

2.2.2. MiRNA in Response to Plant Evolution

MiRNAs control many important aspects of plant development, suggesting that these molecules may also have played key roles in the evolution of developmental processes in plants (Jasinski et al., 2010). The miR164 family in the model angiosperm *A. thaliana* consists of three genes, *ath-miR164a*, *b* and *c*, which are potentially capable of targeting seven members (Gustafson et al., 2005) of the NAC transcription factor family. *Ath-miR164c* mutant shows a slight defect in carpel fusion (Baker et al., 2005), whereas the *ath-miR164abc* triple mutant shows complete carpel separation (Sieber et al., 2007). The expression domains of ath-miR164a, b and c partially overlap (Sieber et al., 2007), which may explain the combination of redundant and non-redundant phenotypes observed in *miR164* mutants. Nikovics and his colleague (2006) found that only *ath-miR164a* shows an increase in the depth of leaf sinuses, whereas both *ath-miR164a* and *ath-miR164b* show an increase in lateral root branching (Guo et al., 2005). The distinct functions of miR164 family members in *A. thaliana* illustrate the processes of gene duplication, gene loss, sub-functionalization and neo-functionalization have clearly also acted on miRNA families, with potentially important consequences for the evolution of plant development. Jasinski et al. (2010) discussed potential roles for *miR164* genes in the evolution of leaf shape and carpel closure in the angiosperms. Evidence for the involvement of the *miR164/CUC* expression balance in carpel closure in multiple ANA grade lineages suggest a potential mechanism for the initial evolution of the closed carpel in early flowering plants.

2.2.3. MiRNA in Response to Biotic and Abiotic Stress

The biotic and abiotic stress causes plants to over- or under-express certain miRNAs or to synthesize new miRNAs to cope with stress. A number of stress-regulated miRNAs have been identified in model plants under various biotic and abiotic stress conditions, including nutrient deficiency, drought, cold, salinity, bacterial infection, UV-B radiation, and mechanical stress (Khraiwesh et al., 2012).

Biotic Stress

Pathogenic bacteria, fungi, viruses, insect pests and nematodes cause severe damage to plants. To combat these pathogenic and parasitic organisms, plants employ a plethora of defense mechanisms. Re-programming of transcriptional cascades is a crucial step in the defense response. Recent discovery revealed that miRNA-guided posttranscriptional regulation plays a crucial role in the plant defense against pathogens through targeting *Transport Inhibitor Response 1* (*TIR1*), an auxin receptor (Navarro et al., 2006). *A. thaliana* miR393 was the first miRNA reported to play a role in plant antibacterial Pattern Triggered Immunity (PTI) by regulating the auxin signaling pathway. MiR393 was induced by a bacterial pattern associated molecular pattern (PAMP) peptide flg22 post-treatment (Navarro et al., 2006). Using small RNA-expression profiling on *A. thaliana* leaves collected at 1 and 3 h post-inoculation, Fahlgren et al. (2007) identified three miRNAs (miR160, miR167, and miR393) that were highly induced and one that was down regulated (miR825) after infection. The role of miRNAs in plant basal defense was further supported by the finding that in *A. thaliana*, miR156 and miR164 were induced by infection with the virus TYMV p69 and were induced in transgenic *A. thaliana* plants expressing the viral silencing suppressor P1/HC-Pro (He et al., 2008). In tobacco, miR156, miR160, and miR164 were induced after virus infection. Upon virus infection, new miRNAs (miR158 and miR1885) were identified in *Brassica*; miR158 and miR1885 were specifically induced by turnip mosaic virus but not by cucumber mosaic virus or tobacco mosaic virus (He et al., 2008). More recent data indicates that miRNAs may affect the response of wheat (*Triticum aestivum* L.) to powdery mildew infection where some of the conserved miRNAs showed differential expression patterns in response to powdery mildew infection: *miR156*, *miR159*, *miR164*, *miR171*, and *miR396*, were downregulated while *miR393*, *miR444*, and *miR827*, were upregulated, respectively (Xin et al., 2010).

Abiotic Stress

More than 42 miRNA families have been reported to be associated with abiotic stress in plants (Sunkar, 2010). Most of these stress-induced miRNAs are evolutionarily conserved, which suggests that miRNAs-mediated regulatory mechanism may be evolutionarily conserved for corresponding environmental stresses in plants. However, the same miRNAs reported to respond abiotic stress in one certain species may not have the same function in other species (Sun et al., 2012).

miRNA in Response to Nutrient Deficiency

Being sessile, plants have to cope with many adverse environmental changes, including changing nutrient availability. Over the last few years, some nutrient-responsive miRNAs have been discovered primarily in the model plant species *A. thaliana*. Recently, more and more miRNA from different plant species was discovered to be induced in response to phosphate-, sulfate-, and nitrogen-deprived conditions, or plant developmental adaptations to metal stress.

Plants have evolved a series of adaptive responses to maintain inorganic phosphate (Pi) homeostasis. Cloning from miRNA libraries and recent efforts in deep sequencing of miRNAs led to the identification of P deficiency-induced miRNAs: miR156, miR399, miR399*, miR2111, miR778, miR827, miR319, miR167, miR157, miR2118, miR159.2, while the expression of miR169, miR395, miR1508, miR1524, miR1532, miR1526 was found to be repressed. The expression of *miR398* was found to be repressed under phosphate deficiency in *Arabidopsis*, while it was up-regulated in nodule of *Phaseolus vulgaris* (*P. vulgaris*). Among them, *miR399*, which targets a gene encoding a putative ubiquitin-conjugating enzyme (UBC or PHO2), is the most thoroughly studied. The crucial regulatory function of *miR399* in the maintenance of Pi homeostasis for plants to adapt to low P growth conditions was reported in many plants, such as *Arabidopsis* (Lin et al., 2008; Fujii et al., 2005; Bari et al., 2006; Chiou et al., 2006; Shukla et al., 2008; Yamasaki et al., 2007), rapeseed (*Brassica napus*) (*B. napus*, Buhtz et al., 2008), pumpkin (*Cucurbita maxima*) (Pant et al., 2008), common bean (*P. vulgaris*) (Valde´s-Lo´pez et al., 2008; 2010), *Medicago truncatula* (*M. truncatula*, Valde´s-Lo´pez et al., 2010). These reports reveal a crucial role of interaction between *miR399* and target gene (*UBC* or *PHO2*) in the maintenance of Pi homeostasis.

This regulation appears to be conserved among different plant species because of the conservation of *miR399* and its target. For example, *miR399* is up-regulated in low phosphate-stressed, whereas its target, putative *UBC* or *PHO2* mRNA, was reduced primarily in roots of plants exposed to low-phosphate stress in *Arabidopsis* (Fujii et al., 2005; Bari et al., 2006; Chiou et al., 2006). Similarly, the expression of *UBC* or *PHO2* is down-regulated in the leaves and roots under phosphate deficiency in common bean (*P. vulgaris*) (Valde´s-Lo´pez et al., 2008), where *miR399* was found to be up-regulated. This is completely in agreement with results from *Arabidopsis* (Bari et al., 2006; Chiou et al., 2006). The putative miR399 target gene (*UBC* or *PHO2*) in *M. truncatula* showed the down-regulation condition under P deficiency conditions (Valde´s-Lo´pez et al., 2010). However, Baek et al. (2013) reported that positive regulation of *miR399* gene expression in response to Pi starvation is mediated at least in part by the transcription factor *AtMYB2*.

MiR395 was also shown to be responsive to P status (Hsieh et al., 2009), being repressed during P limitation, whereas the target *APS4* and *SULTR2;1* transcripts increased in these conditions. It was suggested that suppression of *miR395* during P-limitation may contribute to this upregulation and increase sulfate assimilation and translocation for sulfolipid biosynthesis, thus compensating for the reduced phospholipids during P limitation (Hsieh et al., 2009).

Recently, deep sequencing and reverse transcriptase polymerase chain reaction (RT-PCR) approaches have revealed additional P starvation-responsive miRNAs in *Arabidopsis* except for miR399 and miR395 (Hsieh et al., 2009; Pant et al., 2009). These studies reported the induced expression of *miR156*, *miR778*, *miR827*, *miR2111*, and *miR399** during P starvation, while the expression of *miR169*, *miR395*, and *miR398* was found to be repressed. Hackenberg et al. (2012) found that 55 out of 221 miRNAs were only present in P-deficient treatment while 32 existed only in P-sufficient barley. In total 47 miRNAs had significant differential expression between the two P treatments. In addition, six miRNAs of *P. vulgaris* (*pvu-miR319*, *pvu-miR398*, *pvu-miR167*, *pvu-miR157*, *pvu-miR2118*, *pvu-miR159.2*) were up-regulated under phosphate deficiency in nodule, while four miRNAs from *Glycine max* (*gma-miR1508*, *gma-miR1524*, *gma-miR1532*, *gma-miR1526*) and one from *P. vulgaris* (pvu-miR1511) were both down-regulated under phosphate deficiency in nodule. While *pvu-miR1511*, *gma-miR1524*, *gma-miR1532*, *gma-miR1526*, *pvu-miR167* were up-regulated under phosphate deficiency in roots (Valde´s-Lo´pez et al., 2010). These partially reflected that the same miRNAs reported to respond abiotic stress in a tissue may not have the same function in other tissues of the same species..

The expression of miRNAs shows different responses to sulfate deficiency in plants (Lappartient and Touraine, 1996; Lappartient et al., 1999; Huang et al., 2010). Until now, the unique miRNA described to be involved in sulfate starvation and metabolism is miR395 which targets a low-affinity sulfate transporter SULTR (also called AST68 (At5g10180)) and 3 enzymes in the sulfate assimilation pathway, ATP sulfurylases (APS1: At3g22890; APS3: At4g14680; APS4: At5g43780) (Jones-Rhoades et al., 2006; Sunkar et al., 2007).

To our knowledge, most studies on sulfate starvation were conducted in *A. thaliana* (Jones-Rhoades et al., 2004; Kawashima et al., 2009; Liang et al., 2010; Liang and Yu, 2010) and few on other plants, such as rapeseed (*B. napus*) (Huang et al., 2010; Buhtz et al., 2010). Similar to *miR399/PHO2*, the *miR395/APS-SULTR* regulatory module also appears to be conserved in plants (Huang et al., 2010; Liang et al., 2010). Gene families encoding miR395 were reported from many higher plant species (http://www. mirbase.org), ranging between 2 (tomato) and 24 (rice) known loci. MiR395 family includes six genes, *miR395a-f*, in *A. thaliana*. The six primary *miR395* gene transcripts and the two processed mature miR395 species, that differ by one nucleotide, are strongly induced in low S conditions (Bari et al., 2006; Jones-Rhoades and Bartel, 2004; Pant et al., 2009), although to different levels and in different patterns. The observations that the expression of *miR395* depends on sulfate concentration and that *APS1* mRNA expression declines with increasing *miR395* in *A. thaliana* corroborate the idea that this miRNA regulates sulfate metabolism (Jones-Rhoades et al., 2004). Similarly, increased ATP sulfurylase activity has been reported under sulfur-deprived conditions in *B. napus* and *A. thaliana* (Lappartient et al., 1999; Lappartient and Touraine, 1996). Furthermore, the level of *AST68* mRNA targeted by *miR395* was increased several-fold, especially in the roots of sulfate-deprived *A. thaliana* plants but almost unaffected in leaves (Takahashi et al., 1997).

More details about the function of miR395 in the regulation of sulfate assimilation and translocation have been recently obtained. MiR395 controls the sulfate concentration in the shoot by targeting three *ATP sulfurylase* genes (*APS*), which encode the first enzymes catalyzing sulfate activation in sulfur assimilation pathway. Furthermore, *miR395* also regulates the transport of sulfate between leaves. Under sulfate starvation conditions, up regulated *miR395* represses the expression of *SULTR2;1*, which then confined the transport of sulfate from mature to young leaves. Of note, unlike *APS1* and *APS4* mRNA, *APS3* and shoot *SULTR2;1* is in accordance with *miR395* in response to sulfate deprivation (Liang et al., 2010; Scheible et al., 2011). Down regulation of *APS* genes during S limitation was observed in rapeseed (Huang et al., 2010). These suggest that the differential regulation of targets by *miR395* may be required for adaptation to the sulfate deficiency environment. In addition, they found that there is reciprocal regulation between *SULTR2;1* and *APS* genes through *miR395*.

MiR398 and three other miRNAs (*miR397*, *miR408*, and *miR857*) are reported to be responded to copper (Cu) concentrations and have been proposed to be involved in Cu homeostasis under conditions of Cu deficiency. Among them, *miR39*8, which targets genes encoding *Cu/zinc superoxide dismutase* (*CSD1* and *CSD2*), *cytochrome oxidase subunit V* (*COX5b*, At3g15640), *copper chaperone* (*CCS1*), plantacyanin, is the most thoroughly studied (Abdel-Ghany and Pilon, 2008; Yamasaki et al., 2009; Jones-Rhoades and Bartel 2004; Beauclair et al., 2010; Pant et al., 2009). *MiR408* was shown to target *PLANTACYANIN* (*ARPN*), *laccase3* (*LAC3*), *LAC12*, *LAC13*, *miR397 LAC2*, *LAC4*, *LAC17*, and *miR857 LAC7* (Abdel-Ghany and Pilon, 2008). Under Cu deficiency conditions, target *CSD* and *COX5b-1* transcripts are down regulated. This down regulation is mediated by *miR398* which increases strongly in Cu deficiency (Abdel-Ghany et al., 2005; Yamasaki et al., 2007). In *B. napus* under Cu deprivation, *miR398* is up regulated not only in leaf, stem, and root tissue but also in phloem sap (Buhtz et al., 2008). Similar to *miR399* or *miR395*, these results suggest deep conservation of miR398 function in Cu deficiency. Thus, *miR398* appears to be a key factor in the regulation of Cu homeostasis, by repressing target gene expression of Cu proteins during Cu deficiency.

The transcription factor SQUAMOSA promoter binding protein-like7 (SPL7) has been found to be another central regulator of the copper deficiency response in *A. thaliana* (Yamasaki et al., 2009; Bernal et al., 2012) and *Chlamydomonas reinhardtii* (Kropat et al., 2005). Three *miR398* genes are known in *A. thaliana*. SPL7 binds directly to GTAC *cis* elements in the *miR398* gene promoters in vitro, and these motifs are essential and sufficient for the response to Cu deficiency *in vivo* (Yamasaki et al., 2009). As well as *miR398b* and *miR398c*, the expression of *miR397* and *miR408* was up regulated specifically under low copper (Abdel-Ghany and Pilon, 2008). Similarly, Yamasaki et al. (2009) also found the transcription of *miR398b* and *miR398c*, but not *miR398a*, was strongly induced by *SPL7* in response to Cu deficiency. Analyzing the induction of miRNAs in response to copper deficiency in the wild type and *spl7* showed that the expression of *miR397a* and *miR408* was up regulated under low copper in the wild type, as was expression of *miR857* (Yamasaki et al., 2009). Thus, miRNAs such as miR397, miR398b, miR398c, miR408, and miR857 indirectly affect the levels of many copper proteins regulated by *SPL7* and finally lead plant to adapt to Cu deficiency.

Recently, growing evidence has revealed that miRNAs have crucial roles in the modulation of gene expression, and they also function as key regulators in the alleviation of plant nitrogen stresses. MiRNAs such as miR167, miR169, miR399, miR2111, a miR827-like sequence were found to have functional roles in helping plants to cope with nitrogen (N) deficiency in the soil. It was reported that the *A. thaliana miR169* was strongly down-regulated, whereas its targets, NFYA (Nuclear Factor Y, subunit A) family members, were strongly induced by N starvation. Analysis of the expression of *miR169* precursors showed that *miR169a* was substantially down-regulated in both roots and shoots by N starvation. Accumulation of the NFYA family members was suppressed in transgenic *A. thaliana* with constitutive expression of *miR169a*. Transgenic *A. thaliana* plants overexpressing *miR169a* accumulated less N and were more sensitive to N stress than the wild type (Zhao et al., 2011). These results provide evidence that miRNAs have functional roles in helping plants to cope with fluctuations in N availability in the soil. Recently, *miR167* has been associated with lateral root outgrowth in response to N limitation (Gifford et al., 2008). In addition, *miR399* was down-regulated under N deficiency conditions both in leaves and in roots of common bean (Valde´s-Lo´pez et al., 2010).The putative *miR399* target gene (*PHO2/UBC24*) in *M. truncatula* showed the up-regulation under N deficiency conditions (Valde´s-Lo´pez et al., 2010). Furthermore, miR2111, miR169, a miR827-like sequence, and the abundances of several miRNA* strands were found to be strongly dependent on N status in rapeseed (*B. napus*) phloem sap (Pant et al., 2009).

Recent studies have identified a set of conserved and non-conserved miRNAs that are differentially regulated in response to cadmium (Cd) in rice (Ding et al., 2011), *M. truncatula* (Zhou et al., 2012a), *B. napus* (Zhou et al., 2012b), and *A. thaliana* (Herbette et al., 2006). Cd affects the expression of ERF proteins that belong to the APETALA2 (AP2)/ethylene-responsive-element-binding protein (EREBP) family, and MYB4, MYB43, MYB48 and MYB124, WRKY53 (WRKY family), OBF5 which belongs to the bZIP group of transcription factors. Through PCR-based analyses, it was shown that *miR393* and *miR171* are involved in relieving Cd stress by down-regulating their target genes in *B. napus* (Xie et al., 2007; Huang et al., 2010), rice (Ding and Zhou, 2009; Ding et al., 2011), and *M. truncatula* (Zhou et al., 2008), respectively. Additionally, Huang et al. (2010) isolated 19 potential novel miRNAs that were responsive to Cd by conventional sequencing approaches. Very recently, 19 Cd responsive miRNAs were identified and their target genes were predicted in rice, further validating six of them by quantitative real-time PCR (qPCR) and postulating the potential involvement of certain miRNAs in Cd tolerance in rice (Ding et al., 2011). A qRT-PCR-based assay on Cd stress mediation in *M. truncatula* showed six miRNAs displayed altered expression profiles in response to Cd. Among those, *miR393*, *miR171*, miR319, and *miR529* were up-regulated, whereas *miR166* and *miR398* were down-regulated (Zhou et al., 2008). Osa-miR604 participates in the regulation of Cd tolerance through directing degradation of lipid transfer protein (LTP) mRNA (Xie et al., 2007; Zhou et al., 2008), and it was predicted to target wall-associated kinase (WAK) like protein (Li et al., 2008). Analysis of miRNAs and their targets may provide a new insight into understanding of plant responses to heavy metal stress.

Xie et al. (2007) identified 21 miRNAs in *B. napus* and showed that several miRNAs responded to heavy metals. Using the approach of high-throughput sequencing technology, 52 new miRNAs have been profiled from *M. truncatula* seedlings exposed to mercury (Hg), most being differentially regulated by the heavy metal (Zhou et al., 2012b). Of these, *miR160*, *miR166*, *miR319*, *miR393*, and *miR398* responded to mercury, cadmium and aluminum stresses in *M. truncatula* (Zhou et al., 2008). Among the 347 miRNAs, expression of 23 miRNAs was responsive to aluminum (Al^{3+}), and the majority of Al^{3+}-responsive mRNAs was down-regulated. The majority of Al^{3+} responsive miRNAs belonged to the 'rapid-responsive' category, they were responsive to short-term, but not long-term Al^{3+} treatment

(Chen et al., 2012). These results suggest that miRNAs help regulate plant responses to heavy metal stress in addition to other abiotic stresses.

In common bean plants under manganese toxicity, 11 miRNAs (*miR157*, *miR156*, *miR170*, *miR172*, *miR319*, *miR398*, *pvu-miR159.2*, *pvu-miR2118*, *gma-miR1508*, *gma-miR1526*, and *gma-miR1532*) were strongly induced in nodules and another 11 miRNAs (*miR160*, *miR397*, *miR399*, *miR408*, *pvu-miR1509*, *pvu-miR1514a*, *gmamiR1510*, *gma-miR1511*, *gma-miR1513*, *gma-miR1515*, and *gma-miR1516*) were strongly inhibited in leaves or roots (Valde´s-Lo´pez et al., 2010).

Kong and Yang (2010) found that 8 conserved miRNA (*miR159a*, *miR169b*, *miR169c*, *miR172c*, *miR172d*, *miR173*, *miR394a* and *miR394b*) in 5 families were up-regulated during Fe deficiency in *A. thaliana*. Transcriptional analysis using RT-PCR showed that 70.8% (17/24) of the IDE-containing miRNA genes were expressed in response to Fe deficiency. They presented a putative interaction model between protein-coding genes and miRNA genes under Fe deficiency.

MiRNA in Response to Environmental Conditions

Studies with model plant species, such as *A. thaliana*, rice, *Populus tomentosa* (*P. tomentosa*), have revealed a general role for miRNAs in drought, water, salt, heat, cold responses.

Recent studies have shown that the expression of miRNAs is altered in response to drought stress in many plant species, such as switchgrass (*Panicum virgatum*) (*P. virgatum*, Sun et al., 2012), *P. tomentosa* (Ren et al., 2012), cowpea (Barrera-Figueroa et al., 2011), tobacco (Frazier et al., 2011), *Triticum dicoccoides* (Kantar et al., 2011), soybean (Kulcheski et al., 2011), *P. vulgaris* (Arenas-Huertero et al., 2009), rice (Zhao et al., 2007; Zhou et al., 2010), *Solanum tuberosum* (Kwon et al., 2011), peach (*Prunus persica*) (Eldem et al., 2012). These miRNAs are shown as follows.

MiR169g was confirmed by Zhao et al. (2007) as the only member induced by drought among the 17 members of miR169 family and the induction of miR169g was more prominent in roots than in shoots of rice. They suggested that *miR169g* expression may be regulated directly by *dehydration-responsive element* (*CBF/DREBs*). Zhou et al. (2010) reported that among the 30 miRNAs identified in rice as significantly down- or up-regulated under the drought stress, 11 down-regulated miRNAs (*miR170*, *miR172*, *miR397*, *miR408*, *miR529*, *miR896*, *miR1030*, *miR1035*, *miR1050*, *miR1088*, and *miR1126*) and eight up-regulated miRNAs (*miR395*, *miR474*, *miR845*, *miR851*, *miR854*, *miR901*, *miR903*, and *miR1125*) were revealed for the first time to be induced by drought stress in plants, and nine (*miR156*, *miR168*, *miR170*, *miR171*, *miR172*, *miR319*, *miR396*, *miR397*, and *miR408*) showed opposite expression to that observed in drought-stressed *A. thaliana*. In switchgrass (*P. virgatum*), *miR156*, *miR159*, and *miR396* showed up-regulation expression levels under all drought conditions, while *miR167*, *miR169*, and *miR172* displayed down-regulated expression level under all drought conditions (Sun et al., 2012). As for *miR157*, *miR395* and *miR397* under drought stresses exhibited altered expression pattern in a dose-dependent manner. In peach, the expression level of 262 (104 up-regulated, 158 down-regulated) of the 453 miRNAs changed significantly in leaf tissue, whereas 368 (221 up-regulated, 147 down-regulated) of the 465 miRNAs had expression levels that changed significantly in root tissue upon drought stress (Eldem et al., 2012). Significant changes in the expression of 17 conserved miRNA families and nine novel miRNAs were observed in *P. tomentosa* in response to drought stress, and in seven conserved miRNA families and five novel miRNAs in response to flooding stress (Ren et al., 2012).

Several differentially regulated miRNAs have been also identified in salt-stressed plants. In *A. thaliana*, *miR156*, *miR158*, *miR159*, *miR165*, *miR167*, *miR168*, *miR169*, *miR171*, *miR319*, *miR393*, *miR394*, *miR396*, and *miR397* were upregulated in response to salt stress, while the accumulation of *miR398* was decreased (Liu et al., 2008). In maize, miRNAs displayed different activities in the salt response, and miRNAs belonging to the same miRNA family showed the same behavior. Interestingly, 18 miRNAs were found which were only expressed in the salt-tolerant maize line, and 25 miRNAs that showed a delayed regulation pattern in the salt-sensitive line (Ding et al., 2008). Arenas-Huertero et al. (2009) reported that pvu-miR2118 and pvu-miR159.2 were responsive to drought and salinity in common bean. The expression of *miR162* in switchgrass (Sun et al., 2012) was increased as increasing salinity concentrations, in contrast, *miR397* showed decrease in expression level with the increase of salt concentration. However, the expression level of *miR156*, miR157, *miR159*, *miR167*, *miR172*, *miR395*, *miR398* and *miR399* showed down- or up-regulation under different salt concentration. The expression of *miR417* was regulated by salt stress, dehydration stress, or abscisic acid treatment (Jung and Kang, 2007). To examine the biological role of *miR417* in stress responses, the transgenic *A. thaliana* plants that constitutively overexpress *miR417* under the control of cauliflower mosaic virus 35S promoter were generated, and their phenotypes were analyzed under stress conditions. Results showed that seed germination of the transgenic plants was retarded compared with the wild-type plants in the presence of high salt or abscisic acid. These results imply that miR417 plays a role as a negative regulator of seed germination in *A. thaliana* under salt stress conditions (Jung and Kang, 2007).

Cold stress, including chilling and frost, affects crop yields worldwide by causing tissue injury and delayed growth. Chen and his colleague (2012) demonstrated that *Populus* miRNAs play critical roles in the cold stress response. Differential expression analysis showed that 21 miRNAs were down-regulated and nine miRNAs were up-regulated in response to cold stress. Similarly, Lv et al. (2010) identified 18 cold-responsive rice miRNAs using microarrays. In that study, most were down-regulated. Three miRNAs derived from transposable element sequence were clustered within an intron and proved to be cotranscribed with the host gene only under cold stress. Liu et al. (2008) and Lu et al. (2008) used microarray to analysis all known miRNAs expression profiles under cold stress in *A. thaliana* and *Populus*, respectively. Zhou et al. (2008) developed a transcriptome-based approach to annotating cold-inducible miRNAs in plants. Jian et al. (2009) identified several cold-regulated miRNAs by direct cloning and sequencing. In *Brachypodium*, 28 cold response miRNAs were identified by deep sequencing (Zhang et al., 2009). These results show that plant miRNAs can be regulated by cold stress and may function in critical defense systems for cold fitness.

MiRNA in Response to Reactive Oxygen Species Stress

Exposure of plants to abiotic stresses such as drought, temperature extremes, salinity, high light, heavy metals, and nutrient deficiency causes generation of reactive oxygen species (ROS) and their accumulation to toxic concentrations. In *A. thaliana*, abiotic stress-inducible

CSD1 and *CSD2* have been identified as targets of abiotic stress-down-regulated *miR398*. *MiR398* directs the cleavage of *CSD1* and *CSD2* mRNAs under normal conditions, and down-regulation of *miR398* by oxidative stress results in accumulation of *CSD1* and *CSD2* mRNAs. Transgenic *A. thaliana* plants overexpressing *miR398*-resistant *CSD2* (*mCSD2*) showed more tolerance to oxidative stress conditions than *CSD2* transgenic plants. under stress conditions, the negative regulation of *CSD1* and *CSD2* by *miR398* is relieved, which leads to the increased expression of *CSD1* and *CSD2*, factors that play essential roles in defending against highly toxic O^{2-} (Sunkar et al., 2006).

Genome-wide expression studies in model plant species such as *A. thaliana* found that low oxygen (hypoxia) causes widespread changes in 5–10% of the transcriptome (Klok et al., 2002; Branco-Price et al., 2005; Kreuzwieser et al., 2009). Hypoxia treatment on roots resulted in changes to 25 unique miRNAs (19 families), the majority of these increased under hypoxia, indicating that miRNAs are mainly involved in down-regulating gene expression under hypoxia in *A. thaliana* (Moldovan et al., 2010). These indicate that many of these miRNAs may target hypoxia-specific processes.

2.2.4. MiRNA Involved in Signaling Pathway Affects Plant Development Indirectly

Auxin signaling is one of the most intensively studied pathways mediated by miRNAs in plants. Certain miRNAs target members of the Auxin Response Factors (ARF) family involved in auxin homeostasis and signalling and participate in complex regulatory loops at several crucial stages of root development (Khan et al., 2011). Other miRNAs target and restrict the action of various transcription factors that control root-related processes in several species. Several miRNA families, such as *miR160*, *miR164*, *miR167*, and *miR390*, were demonstrated to be involved in indirectly root cap formation, lateral root development, or adventitious rooting through the auxin signal further transduced by their downstream *ARF* targets (Meng et al., 2010). *MiR160* targets *ARF10*, *ARF16*, and *ARF17*, three of the 23 *A. thaliana ARF* genes (Mallory et al., 2005). *ARF6* and *ARF8* are the target genes of *miR167* (Wu et al., 2006; Rhoades et al., 2002; Bartel and Bartel, 2003). The miR160 family was first reported to be essential for root cap formation in *A. thaliana*. *MiR160* exerts its regulatory effect on root tip growth and gravity sensing through the *ARF10*- and *ARF16*-mediated downstream pathways, and both *ARFs* are targeted by *miR160 in vivo*. Although functionally redundant, both *ARF10* and *ARF16* are indispensable for root cap development in *A. thaliana*. The transgenic result demonstrates that overexpression of *miR160* will cause serious root cap defect in rice, although specific downstream *ARF(s)* or other target families need to be validated. In addition, *miR160* was suggested to have a role in promoting lateral root production in *A. thaliana* through the downstream target *ARF16* (Meng et al., 2010). Mallory et al. (2005) demonstrated that disrupting *miR160* regulation of *ARF17* increases *ARF17* mRNA levels, leads to severe developmental abnormalities, including defects in embryonic, root, vegetative, and floral development, and alters *GH3-like* gene expression. These results indicate that *miR160*-directed regulation is critical for the developmental functions of *ARF17* and expose *ARF17* as a possible transcriptional regulator of *GH3-like* early auxin-response genes.

Guo et al. (2005) reported that the miR164 mediated auxin signaling was required for normal lateral root development in *A. thaliana*. The study showed that one *NAC1* (*NAM/ATAF/CUC1*) was targeted by *miR164*, and *NAC1* transduced auxin signal for lateral root emergence. Thus, auxin-induced expression of *miR164* and subsequent mRNA clearance of *NAC1* provide a homeostatic mechanism to down-regulate auxin signals for lateral root development in *A. thaliana*. Mutant of *miR164* lead to the interrupt of auxin signaling pathway in plant, level increasing of *NAC1* mRNA, and more lateral root formation. While over expression of *miR164* down regulated *NAC1* gene and decrease the formation of lateral root (Achard et al., 2004).

MiR165/166 was demonstrated as an important highly conserved polarizing signal to specify adaxial/abaxial leaf polarity in maize. Since the regulation of *HD-ZIP III* gene expression by *miR165/miR166* predates the origin of angiosperm leaves (Timmermans et al., 2004). *MiR166* initially accumulates immediately below the incipient leaf but subsequently in a progressively broader domain including the adaxial side.

The transport inhibitor response 1 (TIR1, At3g62980) and three auxin signaling F-Box proteins (AFB1, At4g03190), AFB2 (At3g26810), and AFB3 (At1g12820) have been identified and validated as targets of *miR393* (Jones-Rhoades and Bartel, 2004; Navarro et al., 2006). Most studies on *miR393*-mediated regulation of auxin signal transduction were carried out in *A. thaliana* (Vidala et al., 2010; Windels and Vazquez, 2011; Robert-Seilaniantz et al., 2011; Chen et al., 2011), with a few on rice (Bian et al., 2012). It was reported that *miR393* contributes to antibacterial resistance by repressing auxin signaling mediated by its downstream targets, the *TIR1*, *AFB2*, and *AFB3* (Jones-Rhoades and Bartel 2004; Navarro et al., 2006). Remarkably, one circumstantial evidence of the potential role of *miR393* in lateral root formation through signal pathway in *A. thaliana* was mentioned. The regulatory pathway *miR393–TIR1–NAC1*,which likely interacts with the *miR164–NAC1* signaling pathway, may participate in lateral root development. More recently, Chen et al. (2011) demonstrated that *miR393* accumulated in response to exogenous IAA treatment, and its induction was due to enhanced *MIR393b* transcription but not *MIR393a*. Overexpression of a *miR393*-resistant form of *TIR1* (*mTIR1*) enhanced auxin sensitivity and led to pleiotropic effects on plant development including inhibition of primary root growth, overproduction of lateral roots, altered leave phenotype and delayed flowering. Furthermore, *miR393* level was increased in *35S:mTIR1* plant, suggesting that *TIR1* promoted the expression of *miR393* by a feedback loop. The interaction between *miR393* and its target indicates a fine adjustment to the roles of the *miR393-TIR1* module, which is required for auxin responses in plant development.

Additionally, Robert-Seilaniantz et al. (2011) also reported *miR393* redirects secondary metabolite biosynthesis to fine-tune plant defences through repressing auxin signalling, preventing it from antagonizing salicylic acid signalling. They proposed that the action of *miR393* on auxin signalling triggers two complementary responses. First, it prevents suppression of salicylic acid levels by auxin. Second, it stabilizes *ARF1* and *ARF9* in inactive complexes. As a result, the plant is able to mount a full salicylic acid response and to redirect metabolic flow toward the most effective anti-microbial compounds for biotroph resistance.

Intercellular communication, mediated by diverse molecules, is essential for coordinated development and responses to environmental challenge (Dinant and Suárez-López, 2012; Lehesranta et al., 2010; Van Norman et al., 2011). In the case of miRNAs there are reports suggesting that they could act non-cell autonomously within a short range (De Felippes et al., 2011; Juarez et al., 2004; Schwab et al., 2006), and may act mostly in a cell autonomous manner (Marín-González and Suárez-López, 2012). The first proposals that a miRNA can translocate to adjacent cells came from studies of miR165/166 in maize and *A. thaliana* (Juarez et al., 2004; Kidner and Martienssen, 2004). Other recent studies also strongly support the notion that endogenous miRNAs move intercellularly. For example, it was reported that the members of at least two miRNA families, miR165/166 and miR390, together with miR390- dependent ta-siRNAs, act non-cell autonomously and probably move in maize (Nogueira et al., 2009; Vatén et al., 2011).

Metabolites, hormones, proteins and nucleic acids, including mRNAs and sRNA species, act as systemic signals that provide an efficient exchange of information between tissues (Liu et al., 2009; Melnyk et al., 2011; Brosnan and Voinnet, 2011). The importance of long-range movement of miRNAs is related to the function of miRNAs as negative regulators of genes involved in processes that require long-distance signaling (Buhtz et al., 2010; Chuck and O'Connor, 2010). Buhtz et al. (2010) demonstrated that under nutrient starvation *miR395* can be translocated through graft unions from wild type scions to rootstocks of the miRNA processing *hen1-1* mutant. Translocation of *miR395* led to a down-regulation of one of its targets (*APS4*) in rootstocks, suggesting that *miR395* could potentially act as a long distance information transmitter.

2.2.5. Interaction between miRNAs

Current evidence shows that the miR319 and miR164 regulatory networks interact in several ways. First, a TCP3-SDRX repressor activates *CUC* expression ectopically, and the defects observed in the strongest lines were ameliorated in *CUC1* or *CUC2* mutants (Koyama et al., 2007). On the other hand, it has been suggested that the cotyledon fusions observed in *rTCP4* or *rTCP3* (Koyama et al., 2007; Palatnik et al., 2003) are caused by the repression of *CUC* genes, being miR164a, a direct target of *TCP3*.

Plants expressing a *miR319*-resistant version of *TCP4* have increased *miR396* levels and a reduction in *GRF* expression. *GRF2* is normally expressed in a proximodistal gradient along the longitudinal axis of a developing leaf. This gradient is reduced and shifted toward the base of the leaf in *rTCP4* transgenics in parallel with a decrease in cell proliferation (Rodriguez et al., 2010). These results suggest that at least part of the effects of *rTCP4* on cell proliferation might be caused by the activation of *miR396* and a reduction in GRF activity (Rodriguez et al., 2010).

2.2.6. MiRNAs in Formation of ta-siRNAs

Allen et al. (2005) revealed a new surprising role for miRNAs in formation of ta-siRNAs in plants. Two miRNAs, *miR390* and *miR173*, are required to guide cleavage of TAS precursors and initiate ta-siRNA biogenesis in *A. thaliana*. Rather than functioning as negative regulators, *miR173*- and *miR390*-guided cleavage was shown to set the 21-nucleotide phase for ta-siRNA precursor processing. They proposed a model in which miRNA-guided formation of a 5'or 3' terminus within pre-ta-siRNA transcripts, followed by RNA-dependent RNA polymerase (RDR6)-dependent formation of dsRNA and dicer-like processing, yields phased ta-siRNAs that negatively regulate other genes (Allen et al., 2005;Yoshikawa et al., 2005; Bartel, 2005).

Previously, *miR390* was demonstrated to mediate the novel regulatory pathway *miR390–TAS3–ARF2/ARF3/ARF4* involved in auxin signaling. In *A. thaliana*, *miR390*-directed cleavage can define one end of RNA encoded by 3'side of ta-siRNA (TAS3), which subsequently generates phased ta-siRNAs. ARF2, ARF3, and ARF4 are directly targeted by these ta-siRNAs (Williams et al., 2005), which are involved in auxin-mediated polarity establishment in the development of aerial lateral organs, such as leaf and flower. In maize, it was reported that *miR390*, an upstream determinant in leaf polarity whose activity triggers *tas3* ta-siRNA biogenesis, accumulates adaxially in leaves (Nogueira et al., 2009).

3. SiRNA

3.1. SiRNA Biogenesis

The first observation of siRNAs was in tobacco (*Nicotiana tabacum*) and tomato (*Lycopersicon esculentum*) (Hamilton and Baulcombe, 1999). Endogenous siRNAs are processed from long double stranded RNA (dsRNA). The endogenous sources of dsRNAs are: (1) miRNA-directed cleavage products of noncoding transcripts, which are then converted into dsRNAs by RDRs; (2) dsRNAs formed from the mRNAs encoded by natural *cis*-antisense gene pairs (Borsani et al., 2005); and (3) dsRNAs generated from heterochromatin and DNA repeats (Mallory and Vaucheret, 2006). At least three types of endogenous siRNAs exist in plants.

The siRNAs produced by miRNA-directed cleavage of mRNAs are referred to as trans-acting siRNAs (ta-siRNAs). Ta-siRNAs form a class of 21-nt regulatory sRNAs were found only in plants. Ta-siRNAs are generated from non-coding RNA precursors that are initially targeted for cleavage by a miRNA. RDR6 and the DCL enzyme DCL4 act on the cleavage products and convert them into dsRNAs that are cleaved into 21-nt ta-siRNAs (Allen et al., 2005; Yoshikawa et al., 2005; Mallory and Vaucheret, 2006). Ta-siRNAs deriving from the positive and negative strands act in *trans* by guiding the cleavage of endogenous mRNAs, similar to miRNAs. Ta-siRNAs deriving from the negative strand can also act in *cis* by guiding the cleavage of their precursor RNAs. In this way, they participate in feedback regulation of the pathway (Mallory and Vaucheret, 2006). Thus, the accumulation of ta-siRNAs is dependent on the components of both miRNA (DCL1, AGO1, HYL1 and HEN1) and siRNA pathways (RDR6 and DCL4). Ta-siRNAs can guide cleavage of target mRNAs and regulate gene expression at the post-transcriptional level like plant miRNAs. Similar with miRNAs, particular ta-siRNAs can target multiple members of a gene family (Bartel, 2005).

The siRNAs, the second type, derived from dsRNAs formed from the mRNAs encoded by natural *cis*-antisense gene pairs are called natural antisense transcript-derived siRNAs (nat-siRNAs). A region of the dsRNA, resulting from the transcription of overlapping genes, is recognized and processed in a DCL2-dependent manner to a single, 24-nt nat-siRNA species, which directs the DCL1-dependent cleavage of additional 21-nt nat-siRNAs (Mallory and Vaucheret, 2006; Borsani et al., 2005). Although dsRNAs are assumed to result from annealing *cis*-antisense RNA pairs, RDR6 is required for nat-siRNA accumulation. So far, nat-siRNAs have been shown to act only in *cis* by guiding the cleavage of one of the two mRNAs in the *cis*-antisense pair.

The third type of endogenous siRNAs corresponds to heterochromatin and DNA repeats (he-siRNA), such as transposons. The production of these primarily 24-nt siRNAs requires RDR2 and DCL3, and the resulting siRNAs guide chromatin remodeling and, in some cases, DNA methylation at the loci from which they derive (Xie et al., 2004; Chan et al., 2004; Lippman et al., 2004). These modifications are thought to maintain the loci in a transcriptionally silenced state (Mallory and Vaucheret, 2006).

Katiyar-Agarwal et al. (2007) discovered a novel class siRNA recently, long siRNAs (ls-iRNAs), which are 30–40 nt and share many common features with known siRNAs. Generation of AtlsiRNA-1 requires DCL1, DCL4, and the AGO subfamily protein AGO7. It also depends on HYL1, HEN1, HST1, RDR6, and Pol IV (Katiyar-Agarwal et al., 2007).

SiRNA biogenesis has been extensively reviewed. RDRs and DCL proteins process the dsRNAs formed from different sources to produces 21–25 bp siRNAs. The biogenesis of different classes of siRNAs is carried out by specific RDR–DCL combinations (Mallory and Vaucheret, 2006; Sunkar et al., 2007). Subsequently the siRNA/DCL complex is converted to an active RISC containing the mRNA anti-sense siRNA (the guide) strand and an AGO protein. Only one strand of the siRNA will be chosen by the RISC loading complex (RLC) as the guide strand. It is generally accepted that the strand with greater thermodynamic instability at the 5'end is chosen as the guide strand (Schwarz et al., 2003; Khvorova et al., 2003; Bramsen et al., 2009). Hence, the thermodynamic properties of the guide and passenger strands can be modified via base mismatches or by choosing sequences with higher G–C content in the passenger strand 5' end (Snead and Rossi, 2010). RISCs subsequently induce transcriptional or post-transcriptional silencing of complementary sequences. In the RNA cleavage pathway active RISC then performs a single endonucleolytic cleavage of the target RNA as directed by the guide siRNA.

3.2. SiRNA Functions

Nat-siRNAs and Salt Stress

Recent genome-wide analyses have revealed the widespread existence of natural antisense transcript (NATs) in eukaryotic genomes. Work on the founding member of nat-siRNAs, which is derived from a *cis*-NAT gene pair of *SRO5* (*At5g62520*) and *Δ^1-pyrroline-5-carboxylate dehydrogenase* (*P5CDH, At5g62530*) genes, demonstrated an important role of nat-siRNAs in osmoprotection and oxidative stress management under salt stress in *A. thaliana* (Borsani et al., 2005). High salt stress causes accumulation of H_2O_2, and both salt and H_2O_2 induce the expression of *SRO5*. *SRO5* mutant plants also showed hypersensitivity to ROS (H_2O_2). These findings suggest that ROS detoxification under salt stress is mediated by the SRO5 protein. Thus, the SRO5-P5CDH nat-siRNAs together with the P5CDH and SRO5 proteins form an important regulatory loop controlling proline and reactive oxygen species (ROS) production and stress tolerance (Borsani et al., 2005).

Ta-siRNA and Hypoxia Stress

The ta-siRNAs, which belonged to three of the four known ta-siRNA families, TAS1, TAS2, and TAS3, were found to change significantly under hypoxia stress and showed increased abundance under hypoxia in *A. thaliana*. This indicated that ta-siRNAs play a significant and specific role in plant gene regulation and possibly developmental responses to hypoxia (Moldovan et al., 2010).

SiRNA and NaCl, Cold, Heat, and Dehydration Stress

Yao et al. (2010) reported that the expression of four siRNAs changed greatly in the seedlings treated with NaCl (200 mM), cold (4°C for 2h), heat (40°C for 2h), and dehydration stress through northern blot analysis in wheat. Interestingly, both *siRNA 002061_0636_3054.1* and *siRNA 005047_ 0654_1904.1* were down-regulated by heat, NaCl and dehydration stress, while the latter was also up-regulated by cold stress. In addition, *siRNA 080621_1340_ 0098.1* was slightly up-regulated by cold stress and down- regulated by heat but not by NaCl and dehydration stress, while *siRNA 007927_0100_2975.1* was down-regulated by cold, NaCl and dehydration stress but not by heat stress (Yao et al., 2010). This indicated the genes targeted by these siRNAs could play roles in regulating stress responses in wheat. However, the mechanisms of these sRNAs in regulating gene expression and the biological function of target genes need further investigation. The *Craterostigma plantagineum* (*C. plantagineum*) *desiccation tolerant* (*CDT-1*), which is a dehydration-related ABA-inducible gene from the well-known resurrection plant *C. plantagineum*, has been reported to direct the synthesis of a double-stranded 21-bp endogenous siRNA. This siRNA significantly increased the dehydration tolerance thus recover after dehydration of *C. plantagineum* through opening the metabolic pathway for desiccation tolerance (Hilbricht et al., 2008).

SiRNAs and Disease Resistance

The first example of endogenous siRNAs that play a role in bacterial disease resistance in *A. thaliana* was provided by Katiyar-Agarwal and his colleagues (2006). The specific induction of *nat-siRNAATGB2* leads to the silencing of the antisense gene *PPRL*. They reported the *PPRL* is a negative regulator of RPS2 signaling pathway and silencing of *PPRL* by *nat-siRNAATGB2* plays a positive role in disease resistance (Katiyar-Agarwal et al., 2006). They deduced that endogenous siRNA-mediated gene silencing may serve as one important mechanism for gene expression reprogramming in plant defense responses.

The lsiRNAs identified so far are induced by pathogen infection or under specific growth conditions. AtlsiRNA-1 is generated from SRRLK/AtRAP NAT pair and specifically induced by the bacterium Pseudomonas syringae carrying effector avrRpt2. Induction of AtlsiRNA-1 silences AtRAP, which encodes a RAP-domain protein involved in disease resistance (Katiyar-Agarwal et al., 2007).

Ta-siRNA and Developmental Timing and Patterning

Three ta-siRNA gene families (TAS1, TAS2, and TAS3) are known in *A. thaliana*. Biogenesis of *TAS3* ta-siRNAs, which target mRNAs encoding several *ARFs* (including *ARF3/ETTIN* and *ARF4*) involves *miR390*-guided processing of primary transcripts, conversion of a precursor to dsRNA through RDR6 activity, and sequential DCL4-mediated cleavage events. Fahlgren et al. (2006) show that the juvenile-to-adult phase transition is normally suppressed by *TAS3* ta-siRNAs, in an AGO7-dependent manner, through negative regulation of *ARF3* mRNA.

Expression of a nontargeted *ARF3* mutant (*ARF3mut*) in a wild-type background reproduced the phase-change phenotypes detected in *rdr6-15* and *dcl4-2* mutants, which lose all ta-siRNAs. Expression of either *ARF3* or *ARF3mut* in *rdr6-15* plants, in which both endogenous and transgenic copies of *ARF3* were derepressed, resulted in further acceleration of phase change and severe morphological and patterning defects of leaves and floral organs. In light of the functions of *ARF3* and *ARF4* in organ asymmetry, those data reveal multiple roles for *TAS3* ta-siRNA-mediated regulation of *ARF* genes in developmental timing and patterning (Fahlgren et al., 2006).

Ta-siRNA and Leaf and Shoot Development

Adenot et al. (2006) described hypomorphic *rdr6* and *sgs3 Arabidopsis* mutants, which did not exhibit the leaf developmental defects observed in null mutants and which, like null alleles, are impaired in sense-transgene-induced posttranscriptional gene silencing and virus resistance.

Null *rdr6* and *sgs3* mutants lack *TAS1*, *TAS2*, and *TAS3* ta-siRNAs and overaccumulate *ARF3/ETTIN* and *ARF4* mRNAs, which are *TAS3* ta-siRNA targets. A hypomorphic *rdr6* mutant accumulates wild-type *TAS3* ta-siRNA levels but not *TAS1* and *TAS2* ta-siRNAs, suggesting that *TAS3* is required for proper leaf development.

Consistently, *tas3* but not *tas1* or *tas2* mutant exhibits leaf morphology defects, and *ago7/zip* and *drb4* mutants, which exhibit leaf morphology defects, lack *TAS3* but not *TAS1* and *TAS2* ta-siRNAs in leaves. Those results indicate that *TAS3* ta-siRNAs are required for proper leaf development through the action of *AGO7/ZIPPY* (Adenot et al., 2006).

Plants leaves develop proximodistal, dorsoventral (adaxial-abaxial), and mediolateral patterns following initiation. The Myb domain gene *PHANTASTICA* (*PHAN*) is required for adaxial fate in many plants, but the *Arabidopsis* ortholog Asymmetric leaves1 (AS1) has milder effects, suggesting that alternate or redundant pathways exist. Garcia et al. (2006) described enhancers of AS1 with more elongate and dissected leaves. As well as RDR6, an RNA-dependent RNA polymerase previously proposed to influence AS1 through miRNA, these enhancers disrupt *AGO7/ZIPPY*, *SUPPRESSOR OF GENE SILENCING3* (*SGS3*), and *DCL4*, which instead regulate ta-siRNA. Microarray analysis revealed that the ARF genes *ETTIN (ETT)/ARF3* and *ARF4* were upregulated in *AGO7*, whereas *FILAMENTOUS FLOWER (FIL)* was upregulated only in *AS1 AGO7* double mutants. RDR6 and SGS3 likewise repress these genes, which specify abaxial fate. The ta-siRNA gene *TAS3*, which targets *ETT* and *ARF4*, is expressed in the adaxial domain, and *ett AS1 AGO7* triple mutants resemble AS1.

Thus *FIL* is downregulated redundantly by *AS1* and by *TAS3*, acting through *ETT*, revealing a role for ta-siRNA in leaf polarity. RDR6 and DCL4 are required for systemic silencing, perhaps implicating ta-siRNA as a mobile signal (Garcia et al., 2006).

In maize, the polarized expression of sRNAs, both miRNAs and ta-siRNAs, in the incipient leaf directs the patterning of the adaxial-abaxial axis (Nogueira et al., 2007). In rice (*Oryza sativa*), ta-siRNA is considered to be essential for shoot development, including shoot apical meristem (SAM) formation and leaf morphogenesis (Abe et al., 2010).

Ta-siRNA and Auxin Response

All TAS3 transcripts identified to date contain at least one conserved sequence when processed to mature ta-siRNAs that can target ARF family members ARF1, 2, 3, or 4 (tasiR-ARFs). Unlike many miRNAs that have crucial regulatory functions during development and are phylogenetically conserved, only one ta-siRNA, ta-siRNA ARF, is conserved between plant species. It regulates the *ARF3* and *ARF4*, which have important function in development.

In general, processing of angiosperm specific TAS3 loci results in two mature abundant tasiR-ARFs which are nearly identical in sequences and target specificity. ARF1 and 2 contain one tasiR-ARF target site, whereas ARF3 and 4 each contain two conserved target sites and are likely the primary targets of tasiR-ARFs (Axtell et al., 2006; Allen et al., 2005).

Production of two or more nearly identical ta-siRNAs from a locus is a common feature for TAS1 and TAS3, and may be a mechanism to further amplify a suppression signal while maintaining target specificity.

In addition, production of multiple ta-siRNAs could strengthen the suppression gradient developed by ta-siRNAs. When the ta-siRNA source cell is strong, a steeper gradient may be created (Husbands et al., 2009). Studies have shown that ta-siRNAs are mobile, and can travel across multiple cells to create a gradient of suppression activity (Chitwood et al., 2009; Schwab et al., 2009).

Besides small differences in size and their participation in a wide range of siRNA functions, broadly conserved siRNA structure suggests that siRNAs are likely to be functionally equivalent. Thus siRNAs might be expected to retain cross-species function but, for technical reasons, this issue has yet to be rigorously explored (Chau and Lee, 2007).

Conclusion and Future Prospects

Based on the diverse functions of plant miRNAs and siRNAs as reveled above, it is likely that most, if not all, biological processes in plants involve at some point the action of one or more miRNAs /siRNAs. It is discernible that one miRNA may be involved in multiple functions. For example, the known roles of miR393 in nitrate response, in defense against pathogenic bacteria and in plant development lead us to hypothesize that miR393 plays an important role in integrating complex environmental stimuli. Interestingly, many of the miRNA/target genes involved in growth and development are also stress-regulated as well, and many of the miRNA/target genes involved in signaling network also indirectly affect plant growth, development and stress responses. This suggests that the regulatory interaction between miRNAs and their targets are complex with the conclusion raised by Parry et al. (2009) who demonstrated that differences between members of the auxin receptor family may contribute to the complexity of auxin response being a good example of this aspect.

Until now, the biological aspects of few miRNA and siRNA functions have been studied compared with the number of miRNAs registered in miRBase 19.0.. Among these few studies, most results was conducted on model plants and related to tissues such as leaves, flowers, roots, stems. Few results from fruit tissues have been reported owing to most of such studies being based on *Arabidopsis* or other plants which does not have fruit as experimental materials. Many analyses of miRNAs and siRNAs are performed under artificial conditions involving the use of ectopically expressed target transgenes or mutations that confer a gain of function on the miRNA and siRNAs. The real and complex impact of most miRNAs in plant biology networks, however, remains to be ascertained.

In the near future, we anticipate that more plant miRNAs and siRNA will be discovered and numerous mRNAs will be validated as miRNA cleavage targets using various assay detection systems. In the discovery and study of RNAi and related gene-silencing mechanisms, including miRNAs, pioneering work in plants has provided fundamental insights to scientists working with plants. Further studies of development-, other physiological processes-, stress-regulated miRNAs and siRNAs as well as their target genes in plants should identify new components in plant and help to elucidate the complex regulatory network underlying plant development and stress responses. Meanwhile, identification of miRNAs and siRNAs which regulate the stress response or other special characteristics targeted by breeding scientists will help in the design of new strategies for improving stress tolerance or desired characteristics of new plant cultivars.

References

Abe, M., Yoshikawa, T., Nosaka, M., Sakakibara, H., Sato, Y., Nagato, Y., Itoh, J. *WAVY LEAF1*, an ortholog of *Arabidopsis HEN1*, regulates shoot development by maintaining microRNA and trans-acting small interfering RNA accumulation in Rice. *Plant Physiol.* 2010;154: 1335–1346.

Abdel-Ghany, S.E., Müller-Moulé P., Niyogi K.K., Pilon, M., Shikanai, T. Two P-type ATPases are required for copper delivery in *Arabidopsis thaliana* chloroplasts. *Plant Cell* 2005; 17: 1233–1251.

Abdel-Ghany, S.E., Pilon, M. MicroRNA-mediated systemic downregulation of cooper protein expression in response to low copper availability in *Arabidopsis*. *J. Biol. Chem.* 2008; 283: 15932–15945.

Achard, P., Herr, A., Baulcombe, D.C., Harberd, N.P. Modulation of floral development by a gibberellin-regulated microRNA. *Development* 2004; 131: 3357–3365.

Adenot, X., Elmayan, T., Lauressergues, D., Boutet, S., Bouche, N., Gasciolli, V., Vaucheret, H. DRB4-dependent TAS3 trans-acting siRNAs control leaf morphology through AGO7. *Curr. Biol.* 2006; 16: 927–932.

Aida, M., Ishida, T., Fukaki, H., Fujisawa, H. and Tasaka, M. Genes involved in organ separation in *Arabidopsis*: an analysis of the cupshaped cotyledon mutant. *Plant Cell* 1997; 9: 841–857.

Allen, E., Xie, Z.X., Gustafson, A.M. Carrington J.C. MicroRNA-directed phasing during trans-acting siRNA biogenesis in plants. *Cell* 2005; 121: 207–221.

Allen, R.S., Li, J., Stahle, M.I., Dubroue, A., Gubler, F., Millar, A.A. Genetic analysis reveals functional redundancy and the major target genes of the *Arabidopsis* miR159 family. *Proc. Natl. Acad. Sci. U.S.A.* 2007; 104:16371–16376.

Ambros, V., Bartel, B., Bartel, D.P., Burge, C.B., Carrington, J.C., Chen, X., Dreyfuss, G., Eddy, S., Griffiths-Jones, S., Marshall, M., Matzke, M., Ruvkun, G., Tuschl, T. A uniform system for microRNA annotation. *RNA* 2003; 9: 277–279.

Arenas-Huertero, C., Pérez, B., Rabanal, F., Blanco-Melo, D., De la Rosa, C., Estrada-Navarrete, G., Sanchez, F., Covarrubias, A.A., Reyes, J.L. Conserved and novel miRNAs in the legume *Phaseolus vulgaris* in response to stress. *Plant Mol. Biol.* 2009; 70: 385–401.

Aukerman, M.J., Sakai, H. Regulation of flowering time and floral organ identity by a MicroRNA and its *APETALA2-like* target genes. *Plant Cell* 2003; 15: 2730–2741.

Axtell, M., Westholm, J.O., Lai, E.C. Vive la difference: biogenesis and evolution of microRNAs in plants and animals. *Genome Biol.* 2011; 12: 221–234.

Axtell, M.J., Jan, C., Rajagopalan, R., Bartel, D.P. A two-hit trigger for siRNA biogenesis in plants. *Cell* 2006; 127: 565–577.

Baek, D.W., Kim, M.C., Chun, H.J., Kang, S.H., Park, H.C., Shin, G., Park, J.Y., Shen, M.Z., Hong, H. Kim, W.Y., Kim, D.H. Lee, S.Y., Bressan, R.A., Bohnert, H.J., Yun, D.J. Regulation of miR399f transcription by *AtMYB2* affects phosphate starvation responses in *Arabidopsis*. *Plant Physiol.* 2013; 161: 362–373.

Baker, C. C., Sieber, P., Wellmer, F. and Meyerowitz, E. M. The early extra petals1 mutant uncovers a role for microRNA miR164c in regulating petal number in *Arabidopsis*. *Curr. Biol.* 2005; 15: 303–315.

Bari, R., Datt Pant, B., Stitt, M., Scheible, W.R. *PHO2*, microRNA399, and *PHR1* define a phosphate-signaling pathway in plants. *Plant Physiol.* 2006; 141: 988–999.

Barrera-Figueroa, B.E., Gao, L., Diop, N.N., Wu, Z.G., Ehlers, J.D., Roberts, P.A., Close, T.J., Zhu, J.K., Liu, R.Y. Identification and comparative analysis of drought-associated microRNAs in two cowpea genotypes. *BMC Plant Biol.* 2011; 11: 127.

Bartel, B. MicroRNAs directing siRNA biogenesis. *Nat. Struct. Mol. Biol.* 2005; 12: 569-571.

Bartel, B., Bartel, D.P. MicroRNAs: at the root of plant development? *Plant Physiol.* 2003; 132: 709–717.

Bartel, D.P. 2004. MicroRNAs: genomics, biogenesis, mechanism, and function. *Cell* 2004; 116: 281–297.

Baulcombe, D. RNA silencing in plants. *Nature* 2004; 431: 356–363.

Baumberger, N., Baulcombe, D.C. *Arabidopsis* ARGONAUTE1 is an RNA slicer that selectively recruits microRNAs and short interfering RNAs. *Proc .Natl. Acad. Sci. U. S. A.* 2005; 102: 11928–11933.

Beauclair, L., Yu, A., Bouché, N. MicroRNA-directed cleavage and translational repression of the copper chaperone for superoxide dismutase mRNA in Arabidopsis. *Plant J.* 2010; 62: 454–462.

Bernal, M., Casero, D., Singh, V., Wilson, G.T., Grande, A., Yang, H.J., Dodani, S. C., Pellegrini, M., Huijser, P., Connolly, E.L., Merchant, S.S., Kra¨mera, U. Transcriptome sequencing identifies SPL7-regulated copper acquisition genes *FRO4/FRO5* and the copper dependence of iron homeostasis in *Arabidopsis*. *Plant Cell* 2012; 24: 738–761.

Bian, H.W., Xie, Y.K., Guo, F., Han, N., Ma, S.Y., Zeng, Z.H., Wang, J.H., Yang, Y.N., Zhu, M.Y. Distinctive expression patterns and roles of the miRNA393/TIR1 homolog module in regulating flag leaf inclination and primary and crown root growth in rice (*Oryza sativa*). *New Phytol.* 2012; 196: 149–161.

Boualem, A., Laportel, P., Jovanovic, M., Laffont, C., Plet, J., Combier, J.P., Niebel, A., Crespi, M., Frugier, F. MicroRNA166 controls root and nodule development in *Medicago truncatula*. *Plant J.* 2008; 54: 876–887.

Borsani, O., Zhu, J., Verslues, P.E., Sunkar, R., Zhu, J.K. Endogenous siRNAs derived from a pair of natural *cis*-antisense transcripts regulate salt tolerance in *Arabidopsis*. *Cell* 2005; 123: 1279–1291.

Bramsen, J.B., Laursen, M.B., Nielsen, A.F., Hansen, T.B., Bus, C., et al. A large-scale chemical modification screen identifies design rules to generate siRNAs with high activity, high stability and low toxicity. *Nucleic Acids Res.* 2009; 37: 2867–2881.

Branco-Price, C, Kawaguchi, R, Ferreira, R.B., Bailey-Serres, J. Genome-wide analysis of transcript abundance and translation in arabidopsis seedlings subjected to oxygen deprivation. *Annals of Botany* 2005; 96: 647–660.

Brosnan, C.A., Voinnet, O. Cell-to-cell and long-distance siRNA movement in plants: mechanisms and biological implications. *Curr. Opin. Plant Biol.* 2011; 14:580–587.

Buhtz, A., Pieritz, J., Springer, F., Kehr, J. Phloem small RNAs, nutrient stress responses, and systemic mobility. *BMC Plant Biology*. 2010; 10: 64-77.

Buhtz, A., Springer, F., Chappell, L., Baulcombe, D.C., Kehr, J. Identification and characterization of small RNAs from the phloem of *Brassica napus*. *Plant J.* 2008; 53: 739–749.

Chan, S.W., Zilberman, D., Xie, Z., Johansen, L.K., Carrington, J.C., Jacobsen, S.E. RNA silencing genes control de novo DNA methylation. *Science* 2004; 303: 1336.

Chapman, E.J., Carrington, J.C. Specialization and evolution of endogenous small RNA pathways. *Nat. Rev. Genet.* 2007; 8: 884–896.

Chau, B.L., Lee, K.A.W. Function and anatomy of plant siRNA pools derived from hairpin transgenes. *Plant Methods* 2007; 3: 1–14.

Chen, L, Zhang, Y.Y., Ren, Y.Y., Xu, J.C., Zhang, Z.Y., Wang, Y.W. Genome-wide identification of cold-responsive and new microRNAs in *Populus tomentosa* by high-throughput sequencing. *Biochem. Biophys. Res.Commun.* 2012; 417: 892–896.

Chen, X.A. MicroRNA as a translational repressor of *APETALA2* in *Arabidopsis* flower development. *Science* 2004; 303:2022–2025.

Chen, X.M. Small RNAs and their roles in plant development. *Annu. Rev. Cell Dev. Biol.* 2009; 35: 21–44.

Chen, Z.H., Bao, M.L., Sun,Y.Z., Yang, Y.J., Xu, X.H., Wang, J.H., Han,N., Bian, H.W., Zhu, M.Y. Regulation of auxin response by miR393-targeted transport inhibitor response protein 1 is involved in normal development in *Arabidopsis*. *Plant Mol. Biol.* 2011; 77: 619–629.

Chiou, T., Aung, K., Lin, S., Wu, C., Chiang, S., Su, C. Regulation of phosphate homeostasis by microRNA in *Arabidopsis*. *Plant Cell* 2006; 18: 412–421.

Chitwood, D.H., Nogueira, F.T., Howell, M.D., Montgomery, T.A., Carrington, J.C., Timmermans, M.C. Pattern formation via small RNA mobility. *Genes Dev.* 2009; 23: 549–554.

Chitwood, D.H., Timmermans, M.C. Small RNAs are on the move. *Nature* 2010; 467: 415–419.

Chuck, G., Candela, H., Hake, S. Big impacts by small RNAs in plant development. *Curr. Opin. Plant Biol.*2009; 12: 81–86.

Chuck, G., O'Connor, D. Small RNAs going the distance during plant development. *Curr. Opin. Plant Biol.* 2010; 13: 40–45.

Combier, J.P., Frugier, F., de Billy, F.,Boualem, A., Ei-Yahyaoui, F., Moreau, S., Vernie, T., Ott, T., Gamas, P., Crespi, M., Niebel, A.. *MtHAP2-1* is a key transcriptional regulator of symbiotic nodule development regulated by microRNA169 in *Medicago truncatula*. *Genes. Dev*. 2006; 20:3084–3088.

Crawford, B.C., Nath, U., Carpenter, R., Coen, E.S. *CINCINNATA* controls both cell differentiation and growth in petal lobes and leaves of *Antirrhinum*. *Plant Physiol.* 2004;135: 244–253.

De Carvalho, F., Gheysen, G., Kushnir, S., Van Montagu, M., Inze, D., Castresana, C. Suppression of beta-1,3-glucanase transgene expression in homozygous plants. *EMBO J.* 1992; 11: 2595–2602.

De Felippes, F.F., Ott, F., Weigel, D. Comparative analysis of non-autonomous effects of tasiRNAs and miRNAs in *Arabidopsis thaliana*. *Nucleic Acids Res.* 2011; 39:2880–2889.

Dinant, S., Suárez-López, P. Multitude of long-distance signal moleculesvia phloem, in: Witzany, G., Baluˇska, F. (eds.), Biocommunication of Plants. Springer, Heidelberg. 2012; 89–121.

Ding, D., Zhang, L.F., Wang, H., Liu, Z.J., Zhang, Z.X., Zheng, Y.L. Differential expression of miRNAs in response to salt stress in maize roots. *Annals of Botany* 2008; 103: 29–38.

Ding, Y., Chen, Z., Zhu, C. Microarray-based analysis of cadmium-responsive microRNAs in rice (*Oryza sativa*). *J. Exp. Bot.* 2011; 62: 3563–3573.

Ding, Y.F., Zhu, C. The role of microRNAs in copper and cadmium homeostasis. *Biochem. Biophys. Res. Commun.* 2009; 386: 6-10.

Eldem, V., Akcay, U.C., Ozhuner, E., Bakır, Y., Uranbey, S., Unver, T. Genome-wide identification of miRNAs responsive to drought in peach (*Prunus persica*) by high-throughput deep sequencing. *PLoS ONE* 2012; 7(12): e50298.

Emery, J.F., Floyd, S.K., Alvarez, J., Eshed, Y., Hawker, N.P., Izhaki, A., Baum, S.F., Bowman, J.L. Radial patterning of *Arabidopsis* shoots by class III *HD-ZIP* and *KANADI* genes. *Curr. Biol.* 2003; 13: 1768–1774.

Fahlgren, N., Howell, M.D., Kasschau, K.D., Chapman, E.J., Sullivan, C.M., Cumbie, J.S., Givan, S.A., Law, T.F., Grant, S.R., Dangl, J.L., Carrington, J.C. High-throughput sequencing of *Arabidopsis* microRNAs: evidence for frequent birth and death of miRNA genes. *PLoS ONE* 2007; 2:e219.

Fahlgren, N., Montgomery, T.A., Howell, M.D., Allen, E., Dvorak, S.K., Alexander, A.L., Carrington, J.C. Regulation of AUXIN RESPONSE FACTOR3 by TAS3 tasiRNA affects developmental timing and patterning in *Arabidopsis*. *Curr. Biol.* 2006; 16: 939–944.

Frazier, T.P., Sun, G., Burklew, C.E., Zhang, B. Salt and drought stresses induce the aberrant expression of microRNA genes in tobacco. *Mol Biotechnol.* 2011; 49: 159–165.

Fujii H., Chiou, T.J., Lin, S.I., Aung, K., Zhu, J. K. A miRNA involved in phosphate starvation response in *Arabidopsis*. *Curr. Biol.* 2005; 15: 2038–2043.

Garcia, D. A miRacle in plant development: role of microRNAs in cell differentiation and patterning. *Semin. Cell Dev. Biol.* 2008; 19: 586–595.

Garcia, D., Collier, S.A., Byrne, M.E., Martienssen, R.A. Specification of leaf polarity in *Arabidopsis* via the trans-acting siRNA pathway. *Curr. Biol.* 2006; 16: 933–938.

Gifford, M.L., Dean, A., Gutierrez, R.A., Coruzzi, G.M., Birnbaum, K.D. Cell-specific nitrogen responses mediate developmental plasticity. *Proc. Natl. Acad. Sci. U.S.A.* 2008; 105: 803–808.

Giraud, E., Ng, S., Carrie, C., Duncan, O., Low, J., Lee, C.P., Aken, O.V., Millar, A.H., Murcha, M., Whelan, J. TCP transcription factors link the regulation of genes encoding mitochondrial proteins with the circadian clock in *Arabidopsis thaliana*. *Plant Cell* 2010; 22:3921–3934.

Grigg, S.P., Canales, C., Hay, A., Tsiantis, M. SERRATE coordinates shoot meristem function and leaf axial patterning in *Arabidopsis*. *Nature* 2005; 437: 1022–1026.

Guo, H.S., Xie, Q., Fei, J.F., Chua, N. H. MicroRNA directs mRNA cleavage of the transcription factor NAC1 to downregulate auxin signals for *Arabidopsis* lateral root development. *Plant Cell* 2005; 17: 1376–1386.

Gustafson, A. M., Allen, E., Givan, S., Smith,D., Carrington, J. C., Kasschau, K. D. ASRP: the *Arabidopsis* small RNA project database. *Nucleic Acids Res.* 2005; 33: 637–640.

Hackenberg, M., Huang, P.J., Huang, C.Y., Shi, B.J., Gustafson, P., Langridge, P. A comprehensive expression profile of microRNAs and other classes of non-coding small RNAs in barley under phosphorous-deficient and -sufficient conditions. *DNA Res.* 2012; 1–17.

Hamilton, A.J, Baulcombe, D.C. A species of small antisense RNA in posttranscriptional gene silencing in plants. *Science* 1999; 286: 950-952.

Hammond, S.M., Boettcher, S., Caudy, A.C., Kobayashi, R., Hannon, G.J. Argonaute2, a link between genetic and bio- chemical analyses of RNAi. *Science* 2001; 293: 1146–1150.

Hannon, G.J. RNA interference. *Nature* 2002; 418: 244–251.

He, X.F., Fang, Y.Y., Feng, L., Guo, H.S. Characterization of conserved and novel microRNAs and their targets, including a *TuMV*-induced *TIR-NBS-LRR* class R gene-derived novel miRNA in *Brassica. FEBS Lett.* 2008; 582: 2445–2452.

Herbette, S., Taconnat, L., Hugouvieux, V., Piette, L., Magniette, M.L., Cuine, S., Auroy, P., Richaud, P., Forestier, C., Bourguignon, J., Renou, J.P., Vavasseur, A., Leonhardt, N. Genome-wide transcriptome profiling of the early cadmium response of *Arabidopsis* roots and shoots. *Biochimie.* 2006; 88: 1751–1765.

Hilbricht, T., Varotto, S., Sgaramella, V., Bartels, D., Salamini, F., Furini, A. Retrotransposons and siRNA have a role in the evolution of desiccation tolerance leading to resurrection of the plant *Craterostigma plantagineum. New Phytol.* 2008; 179: 877–887.

Hsieh, L.C., Lin, S.I., Shih, A.C.C., Chen, J.W., Lin, W.Y., Tseng, C.Y., Li, W.H., Chiou, T.J. Uncovering small RNA-mediated responses to phosphate deficiency in *Arabidopsis* by deep sequencing. *Plant Physiol.* 2009; 151: 2120–2132.

Huang, S.Q., Xiang, A.L., Che, L.L., Chen, S., Li, H., Song, J.B., Yang, Z.M. A set of miRNAs from *Brassica napus* in response to sulfate-deficiency and cadmium stress. *Plant Biotechno. J.* 2010; 8: 887–899.

Husbands, A.Y., Chitwood, D.H., Plavskin, Y., Timmermans, M.C. Signals and prepatterns: new insights into organ polarity in plants. *Genes. Dev.* 2009; 23: 1986–1997.

Jasinski,S.,Vialette-Guiraud, A.C.M., Scutt, C.P. The evolutionary-developmental analysis of plant microRNAs. *Phil. Trans. R. Soc. B.* 2010; 365: 469–476.

Jian, X., Zhang, L., Li, G., Zhang, L., Wang, X., Cao, X., Fang, X., Chen, F. Identification of novel stress-regulated microRNAs from *Oryza sativa* L. *Genomics.* 2009; 95: 47–55.

Jones-Rhoades, M.J., Bartel, B., Bartel, D.P. MicroRNAs and their regulatory roles in plants. *Annu. Rev. Plant Biol.* 2006; 57: 19–53.

Jones-Rhoades, M.W., Bartel, D.P. Computational identification of plant microRNAs and their targets, including a stress-induced miRNA. *Mol. Cell* 2004; 14: 787–799.

Juarez, M.T., Kui, J.S., Thomas, J., Heller, B.A., Timmermans, M.C.P. microRNA mediated repression of *rolled leaf1* specifies maize leaf polarity. *Nature* 2004; 428: 84–88.

Jung, H.J., Kang, H. Expression and functional analyses of microRNA417 in *Arabidopsis thaliana* under stress conditions. *Plant Physiol. Biochem.* 2007; 45: 805–811.

Kaneko, M., Inukai, Y., Ueguchi-Tanaka, M., Itoh, H., Izawa, T., Kobayashi, Y., Hattori, T., Miyao, A., Hirochika, H., Ashikari, M., Matsuoka, M. Loss-of-function mutations of the rice *GAMYB* gene impair alpha-amylase expression in aleurone and flower development. *Plant Cell* 2004;16:33–44.

Kantar, M., Lucas, S.J., Budak, H. miRNA expression patterns of Triticum dicoccoides in response to shock drought stress. *Planta* 2011; 233: 471–484.

Katiyar-Agarwal, S., Gao, S., Vivian-Smith, A., Jin, H.L. A novel class of bacteria-induced small RNAs in *Arabidopsis*. *Genes. Dev.* 2007; 21: 3123–3134.

Katiyar-Agarwal, S., Morgan, R., Dahlbeck, D., Borsani ,O., Villegas, Jr A., Zhu, J.K., Staskawicz, B.J., Jin, H. A pathogen-inducible endogenous siRNA in plant immunity. *Proc. Natl. Acad. Sci. U. S. A.* 2006; 103: 18002–18007.

Kawashima, C.G., Yoshimoto, N., Maruyama-Nakashita, A., Tsuchiya, Y.N., Saito, K.,Takahashi, H., Dalmay, T. Sulphur starvation induces the expression of microRNA-395 and one of its target genes but in different cell types. *The Plant J.* 2009; 57: 313–321.

Khan, G.A., Declerck, M., Sorin, C., Hartmann, C., Crespi, M., Lelandais-Brière, C. MicroRNAs as regulators of root development and architecture. *Plant Mol. Biol.* 2011;77:47-58.

Khraiwesh, B., Zhu, J.K., Zhu, J.H. Role of miRNAs and siRNAs in biotic and abiotic stress responses of plants. *Biochim. Biophys. Acta.* 2012; 1819: 137–148.

Khvorova, A., Reynolds, A., Jayasena, S.D. Functional siRNAs and miRNAs exhibit strand bias. *Cell* 2003; 115: 209–216.

Kidner, C.A., Martienssen, R.A. Spatially restricted microRNA directs leaf polarity through ARGONAUTE1. *Nature* 2004; 428: 81–84.

Kim, J., Jung, J.H., Reyes, J.L., Kim, Y.S., Kim, S.Y., Chung, K.S., Kim, J.A., Lee, M.S., Lee, Y., Kim, V.N., Chua, N.H., Park, C.M. MicroRNA-directed cleavage of *ATHB15* mRNA regulates vascular development in Arabidopsis inflorescence stems. *Plant J.* 2005; 42: 84–94.

Kim, S.H., Yang, J.Y., Xu J., Jang, I.C., Prigge, M.J., Chua, N.H. Two cap-binding proteins CBP20 and CBP80 are involved in processing primary microRNAs. *Plant Cell Physiol.* 2008; 49: 1634–1644.

Kim, V.N. MicroRNA precursors in motion: exportin-5 mediates theirnuclear export. *Trends Cell Biol.* 2004; 14: 156–159.

Klok, E.J., Wilson, I.W., Wilson, D., Chapman, S.C., Ewing, R.M., Somerville, S.C., Peacock, W.J., Dolferus, R., Dennis, E.S. Expression profile analysis of the low-oxygen response in *Arabidopsis* root cultures. *The Plant Cell* 2002; 14: 2481–2494.

Kong, W.W., Yang, Z.M. Identification of iron-deficiency responsive microRNA genes and cis-elements in *Arabidopsis*. *Plant Physiol. Biochem.* 2010; 48: 153–159.

Koyama, T., Furutani, M., Tasaka, M., Ohme-Takagi, M. TCP transcription factors control the morphology of shoot lateral organs via negative regulation of the expression of boundary specific genes in *Arabidopsis*. *Plant Cell* 2007; 19: 473–484.

Kozomara, A., Griffiths-Jones, S. MiRBase: integrating microRNA annotation and deep-sequencing data. *Nuclear Acid Res.* 2011; 39: 152–157.

Kreuzwieser, J., Hauberg, J., Howell, K.A., Carroll, A., Rennenberg, H., Millar, A.H., Whelan, J. Differential response of grey poplar leaves and roots underpins stress adaptation during hypoxia. *Plant Physiol.* 2009; 149: 461–473.

Kropat, J., Tottey, S., Birkenbihl, R.P., Depège, N., Huijser, P., Merchant, S. A regulator of nutritional copper signaling in *Chlamydomonas* is an SBP domain protein that recognizes the GTAC core of copper response element. *Proc. Natl. Acad. Sci. U.S.A.* 2005; 102: 18730–18735.

Kulcheski, F.R., de Oliveira, L.F., Molina, L.G., Almerao, M.P., Rodrigues, F.A., Marcolino, J., et al. Identification of novel soybean microRNAs involved in abiotic and biotic stresses. *BMC Genomics* 2011; 12: 307.

Kwon, H.B., Hwang, E.W., Shin, S.J., Park, S.C., Jeong, M.J. Identification of miR172 family members and their putative targets responding to drought stress in *Solanum tuberosum*. *Genes and Genomics* 2011; 33: 105–110.

Lagos-Quintana, M., Rauhut, R., Lendeckel, W., Tuschl, T. Identification of novel genes coding for small expressed RNAs. *Science* 2001; 294: 853– 858.

Lappartient, A.G., Touraine, B. Demand-driven control of root ATP sulfurylase activity and SO42− uptake in intact canola (the role of phloem-translocated glutathione). *Plant Physiol.* 1996; 111: 147–157.

Lappartient, A.G., Vidmar, J.J., Leustek, T., Glass, A.D.M., Touraine, B. Inter-organ signaling in plants: regulation of ATP sulfurylase and sulfate transporter genes expression in roots mediated by phloem-translocated compound. *Plant J.*1999; 18: 89–95.

Lau, N.C., Lim, L.P., Weinstein, E.G., Bartel, D.P. An abundant class of tiny RNAs with probable regulatory roles in *Caenorhabditis elegans*. *Science* 2001; 294: 858–862.

Laufs, P., Peaucelle, A., Morin, H. and Traas, J. MicroRNA regulation of the CUC genes is required for boundary size control in *Arabidopsis* meristems. *Development.* 2004; 131: 4311–4322.

Lauter, N., Kampani, A., Carlson, S., Goebel, M., Moose, S.P. MicroRNA172 downregulates *glossy15* to promote vegetative phase change in maize. *Proc. Natl. Acad. Sci. U. S. A.* 2005; 102: 9412–9417.

Lee, R.C., Ambros, V. An extensive class of small RNAs in *Caenorhabditis elegans*. *Science* 2001; 294: 862– 864.

Lee, R.C., Feinbaum, R.L., Ambros, V. The *C. elegans* heterochronic gene *lin-4* encodes small RNAs with antisense complementarity to *lin-14*. *Cell* 1993; 75: 843–854.

Lehesranta, S.J. R. Lichtenberger, Y. Helariutta, Cell-to-cell communication in vascular morphogenesis. *Curr. Opin. Plant Biol.* 2010; 13: 59–65.

Li, H., Zhou, S.Y., Zhao, W.S., Su, S.C., Peng, Y.L. A novel wall-associated receptorlike protein kinase gene, *OsWAK1*, plays important roles in rice blast disease resistance. *Plant Mol. Biol.* 2008; 69: 337–346.

Liang, G., Yang, F., Yu, D. microRNA395 mediates regulation of sulfate accumulation and allocation in *Arabidopsis thaliana*. *Plant J.* 2010; 62: 1046–1057.

Liang, G., Yu, D.Q. Reciprocal regulation among miR395, APS and SULTR2;1 in *Arabidopsis thaliana*. *Plant Signal. Behav.* 2010; 5:10: 1257–1259.

Lin, S.I., Chiang, S.F., Lin, W.Y., Chen, J.W., Tseng, C.Y., Wu, P.C., Chiou, T.J. Regulatory network of microRNA399 and PHO2 by systemic signaling. *Plant Physiol.* 2008; 147: 732–746.

Lippman, Z., Gendrel, A.V., Black, M., Vaughn, M.W., Dedhia, N., McCombie, W.R., Lavine, K., Mittal, V., May, B., Kasschau, K.D., Carrington, J.C., Doerge, R.W., Colot, V., Martienssen, R. Role of transposable elements in heterochromatin and epigenetic control. *Nature* 2004; 430: 471–476.

Liu, H.H., Tian, X., Li, Y.J., Wu, C.A., Zheng, C.C. Microarray-based analysis of stress-regulated microRNAs in *Arabidopsis thaliana*. *RNA* 2008; 14: 836–843.

Liu, T.Y., Chang, C.Y., Chiou, T.J. The long-distance signaling of mineral macronutrients. *Curr. Opin. Plant Biol.* 2009; 12:312–319.

Llave, C., Kasschau, K.D., Rector, M.A., Carrington, J.C. Endogenous and silencing-associated small RNAs in plants. *Plant Cell* 2002; 14: 1605–1619.

Lu, S., Sun, Y., Chiang, V. Stress-responsive microRNAs in *Populus*. *Plant J.* 2008; 55: 131–151.

Lv, D.K., Bai, X., Li, Y., Ding, X.D., Ge, Y., Cai, H., Jia, W, Wu, N., Zhu, Y.M. Profiling of cold-stress-responsive miRNAs in rice by microarrays. *Gene* 2010; 459: 39–47.

Mallory, A. C., Vaucheret, H. Functions of microRNAs and related small RNAs in plants. *Nat. Genet.* 2006; 38: 31–36.

Mallory, A.C., Bartel, D.P., Bartel, B. MicroRNA-directed regulation of *Arabidopsis* AUXIN RESPONSE FACTOR17 is essential for proper development and modulates expression of early auxin response genes. *Plant Cell* 2005; 17: 1360–1375.

Mallory, A.C., Dugas, D.V., Bartel, D.P. and Bartel, B. MicroRNA regulation of NAC-domain targets is required for proper formation and separation of adjacent embryonic, vegetative, and floral organs. *Curr. Biol.* 2004b; 14: 1035–1046.

Mallory, A.C., Reinhart, B.J., Jones-Rhoades, M.W., Tang, G., Zamore, P.D., Barton, M.K., Bartel, D.P. MicroRNA control of *PHABULOSA* in leaf development: importance of pairing to the microRNA 50 region. *EMBO J.* 2004a; 23: 3356–3364.

Marín-González, E., Suárez-López, P. "And yet it moves": cell-to-cell and long-distance signaling by plant microRNAs. *Plant Sci.* 2012; 196: 18–30.

McConnell, J., Emery, J., Eshed, Y., Bao, N., Bowman, J., Barton, M.K. Role of PHABULOSA and PHAVOLUTA in determining radial patterning in shoots. *Nature* 2001; 411: 709–713.

Melnyk, C.W., Molnar, A., Baulcombe, D.C. Intercellular and systemic move-ment of RNA silencing signals. *EMBO J.*2011; 30: 3553–3563.

Meng, Y.J., Ma, X.X., Chen, D.J., Wu, P., Chen, M. MicroRNA-mediated signaling involved in plant root development. *Biochem. Biophys. Res. Commun.* 2010; 393: 345–349.

Mette, M.F., Van der Winden J., Matzke M., Matzke A.J. Short RNAs can identify new candidate transposable element families in *Arabidopsis*. *Plant Physiol.* 2002; 130: 6–9.

Millar, A.A., Gubler, F. The *Arabidopsis* GAMYB-like genes, *MYB33* and *MYB65*, are microRNA-regulated genes that redundantly facilitate anther development. *Plant Cell* 2005; 17:705–721.

Mlotshwa, S., Yang, Z., Kim, Y., Chen, X. Floral patterning defects induced by *Arabidopsis APETALA2* and microRNA172 expression in *Nicotiana benthamiana*. *Plant Mol. Biol.* 2006; 61:781-93.

Moldovan, D., Spriggs, A., Yang, J., Pogson, B.J., Dennis, E.S., Wilson, I.W. 2010. Hypoxia-responsive microRNAs and trans-acting small interfering RNAs in *Arabidopsis*. *J. Exp. Bot.* 2010; 61: 165–177.

Nag, A., King, S., Jack, T. MiR319a targeting of TCP4 is critical for petal growth and development in *Arabidopsis*. *Proc. Natl. Acad. Sci. U.S.A.* 2009; 106: 22534–22539.

Napoli, C., Lemieux, C., Jorgensen, R. Introduction of a chimeric chalcone synthase gene into petunia results in reversible co-suppression of homologous genes intrans. *Plant Cell* 1990; 2: 279–289.

Navarro, L., Dunoyer P., Jay F., Arnold B., Dharmasiri N., Estelle M., Voinnet O., Jones J.D.G. A plant miRNA contributes to antibacterial resistance by repressing auxin signaling. *Science* 2006; 312: 436–439.

Nikovics, K., Blein, T., Peaucelle, A., Ishida, T., Morin, H., Aida, M., Laufs, P. The balance between the miR164A and CUC2 genes controls leaf margin serration in *Arabidopsis. Plant Cell* 2006; 18: 2929–2945.

Nogueira, F.T. S., Chitwood, D.H., Madi, S., Ohtsu, K., Schnable, P.S., Scanlon, M.J., Timmermans, M.C.P. Regulation of small RNA accumulation in the maize shoot apex. *PLoS Genet.* 2009; 5: e1000320.

Nogueira, F.T.S., Madi, S., Chitwood, D.H., Juarez, M.T., Timmermans, M.C.P. Two small regulatory RNAs establish opposing fates of a developmental axis. *Genes. Dev.* 2007; 21: 750–755.

Ori, N., Cohen, A.R., Etzioni, A., Brand, A., Yanai, O., Shleizer, S., et al. Regulation of *LANCEOLATE* by miR319 is required for compound-leaf development in tomato. *Nat. Genet.* 2007; 39: 787–791.

Palatnik, J.F., Allen, E., Wu, X., Schommer, C., Schwab, R., Carrington, J.C., Weigel, D. Control of leaf morphogenesis by microRNAs. *Nature* 2003; 425: 257–263.

Palatnik, J.F., Wollmann, H., Schommer, C., Schwab, R., Boisbouvier, J., Rodriguez, R., Warthmann, N., Allen, E., Dezulian, T., Huson, D., Carrington, J.C., Weigel, D. Sequence and expression differences underlie functional specialization of *Arabidopsis* microRNAs miR159 and miR319. *Dev. Cell* 2007; 13: 115–125.

Pant, B.D., Buhtz, A., Kehr, J., Scheible, W.R. MicroRNA399 is a longdistance signal for the regulation of plant phosphate homeostasis. *Plant J.* 2008; 53: 731–738.

Pant, B.D., Musialak-Lange, M., Nuc, P., May, P., Buhtz, A., Kehr, J., Walther, D., Scheinle, W.R. Identification of nutrient-responsive *Arabidopsis* and rapeseed micorRNAs by comprehensive real-timepolymerase chain reaction profiling and small RNA sequencing. *Plant Physiol.* 2009; 150: 1541–1555.

Park, M.Y., Wu, G., Gonzalez-Sulser, A., Vaucheret, H., Poethig, R.S. Nuclear processing and export of microRNAs in *Arabidopsis. Proc. Natl. Acad. Sci. U.S.A.* 2005; 102: 3691–3696.

Park, W., Li, J., Song, R., Messing, J., Chen, X. CARPEL FACTORY, a Dicer homolog, and HEN1, a novel protein, act in microRNA metabolism in *Arabidopsis thaliana. Curr. Biol.* 2002; 12: 1484–1495.

Parker, J.S., Mark Roe S., Barford, D. Structural insights into mRNA recognition from a PIWI domain–siRNA guide complex. *Nature* 2005; 434: 663-666.

Parry, G., Calderon-Villalobos, L. I., Prigge, M., Peret, B., Dharmasiri, S., Itoh, H., Lechner, E., Gray, W. M., Bennett, M., Estellea, M. Complex regulation of the TIR1/AFB family of auxin receptors. *Proc. Natl. Acad. Sci. U. S. A.* 2009; 106: 22540–22545.

Pedersen, I., David, M. MicroRNAs in the immune response. *Cytokine* 2008; 43: 391–394.

Raman, S., Greb, T., Peaucelle, A., Blein, T., Laufs, P., Theres, K. Interplay of miR164, *CUP-SHAPED COTYLEDON* genes and *LATERAL SUPPRESSOR* controls axillary meristem formation in *Arabidopsis thaliana. Plant J.* 2008;55: 65–76.

Rand, T. A., Ginalski, K., Grishin, N.V., Wang, X.D. Biochemical identification of Argonaute 2 as the sole protein required for RNA-induced silencing complex activity. *Proc. Natl. Acad. Sci. USA.* 2004;101: 14385–14389.

Reinhart, B.J., Weinstein, E.G., Rhoades, M.W., Bartel, B., Bartel, D.P. MicroRNAs in plants. *Genes. Dev.* 2002; 16: 1616–1626.

Ren, Y., Chen, L., Zhang, Y., Kang, X., Zhang, Z., Wang, Y. Identification of novel and conserved *Populus tomentosa* microRNA as components of a response to water stress. *Funct. Integr. Genomics*. 2012; 12: 327-339.

Rhoades, M.W., Reinhart, B.J., Lim, L.P., Burge C.B., Bartel, B., Bartel1, D.P. Prediction of plant microRNA targets. *Cell* 2002; 110: 513–520.

Robert-Seilaniantz, A., MacLean, D., Jikumaru, Y., Hill, L., Yamaguchi, S.., Kamiya, Y., Jones, J. D. G. The microRNA miR393 re-directs secondary metabolite biosynthesis away from camalexin and towards glucosinolates. *Plant J.* 2011; 67: 218–231.

Rodriguez, R.E., Mecchia, M.A., Debernardi, J.M., Schommer, C., Weigel, D., Palatnik, J.F. Control of cell proliferation in *Arabidopsis thaliana* by microRNA miR396. *Development* 2010;137: 103–112.

Romano, N., Macino, G. Quelling: transient inactivation of gene expression in *Neurospora crassa* by transformation with homologous sequences. *Mol. Microbiol.* 1992; 6: 3343–3353.

Sarvepalli, K., Nath, U. Hyper-activation of the TCP4 transcription factor in *Arabidopsis thaliana* accelerates multiple aspects of plant maturation. *Plant J.* 2011; 67: 595–607.

Scheible, W.R., Pant, B.D., Musialak-Lange, M., Nuc, P. Nutrient-responsive plant microRNAs, in Non Coding RNAs in Plants. Erdmann V.A. and Barciszewski J. (eds.) 2011; 317-337.

Schommer C., Bresso, E.G., Spinelli, S.V., Palatnik, J.F. Role of microRNA miR319 in plant development, in: Sunkar R. (ed.), MicroRNAs in Plant Development and Stress Responses. 2012; P:296.

Schommer, C., Palatnik, J.F., Aggarwal, P., Chetelat, A., Cubas, P., Farmer, E.E., Nath, U., Weigel, D. Control of jasmonate biosynthesis and senescence by miR319 targets. *PLoS Biol.* 2008; 6(9): e230.

Schwab, R., Maizel, A., Ruiz-Ferrer, V., Garcia, D., Bayer, M., Crespi, M. Endogenous TasiRNAs mediate non-cell autonomous effects on gene regulation in *Arabidopsis thaliana*. *PLoS ONE* 2009; 4:e5980.

Schwab, R., Ossowski, S., Riester, M., Warthmann, N., Weigel, D. Highly specific gene silencing by artificial microRNAs in *Arabidopsis*. *Plant Cell* 2006; 18: 1121–1133.

Schwab, R., Palatnik, J.F., Riester, M., Schommer, C., Schmid, M., Weigel, D. Specific effects of microRNAs on the plant transcriptome. *Dev. Cell* 2005; 8: 517–527.

Schwarz, D.S., Hutvagner, G., Du, T., Xu, Z., Aronin, N., Zamore, P.D. Asymmetry in the assembly of the RNAi enzyme complex. *Cell* 2003; 115: 199–208.

Shukla L.I., Chinnusamy V., Sunkar R. The role of microRNAs and other endogenous small RNAs in plant stress responses. *Biochim. Biophys. Acta.* 2008; 1779:743–748.

Sieber, P., Wellmer, F., Gheyselinck, J., Riechmann, J.L., Meyerowitz, E.M. Redundancy and specialization among plant microRNAs: role of the MIR164 family in developmental robustness. *Development* 2007;134: 1051–1060.

Snead, N.M., Rossi, J.J. Biogenesis and function of endogenous and exogenous siRNAs. *Wiley. Interdiscip. Rev. RNA.* 2010; 1:117-131.

Sun, G., Stewart, C.N.Jr., Xiao, P., Zhang, B.H. MicroRNA expression analysis in the cellulosic biofuel crop switchgrass (*Panicum virgatum*) under abiotic stress. *PLoS ONE* 2012; 7: e32017.

Sunkar R., Chinnusamy V., Zhu J., Zhu J.K. Small RNAs as big players in plant abiotic stress responses and nutrient deprivation. *Trends Plant Sci*. 2007; 12: 301–309.

Sunkar, R. MicroRNA in plant development and stres responses: Carla Schommer, Edgardo G. Bresso, Silvana V. Spinelli, and Javier F. Palatnik. Role of microRNA miR319 in plant development. Springer-Verlag Berlin Heidelberg. Argentina. 2012; ISBN: 978-3-642-27383-4.

Sunkar, R. MicroRNAs with macro-effects on plant stress responses. *Seminars in Cell and Developmental Biology* 2010; 21: 805–811.

Sunkar, R., Kapoor, A., Zhu, J.K. Posttranscriptional induction of two Cu/Zn superoxide dismutase genes in *Arabidopsis* is mediated by downregulation of miR398 and important for oxidative stress tolerance. *Plant Cell* 2006; 18:2051–2065.

Takahashi, H., Yamazaki, M., Sasakura, N., Watanabe, A., Leustek, T., Engler, J.D., Engler, G., Van Montagu, M., Saito, K. Regulation of sulfur assimilation in higher plants: a sulfate transporter induced in sulfate-starved roots plays a central role in *Arabidopsis thaliana*. *Proc. Natl. Acad. Sci. U. S. A.* 1997; 94: 11102–11107.

Timmermans, M.C.P., Juarez, M.T., Phelps-durr T.L. A conserved microRNA signal specifies leaf polarity. *Cold Spring Harb. Symp. Quant. Biol.* 2004; 69: 409-418.

Valde´s-Lo´pez, O., Yang, S. S., Aparicio-Fabre, R., Graham, P.H., Reyes, J.L., Vance, C.P., Herna´ndez, G. MicroRNA expression profile in common bean (*Phaseolus vulgaris*) under nutrient deficiency stresses and manganese toxicity. *New Phytol.* 2010; 187: 805–818.

Valde´s-Lo´pez, O., Arenas-Huertero, C., Ramı´rez, M., Girard, L., Sa´nchez, F.,Vance, C.P., Reyes, J.L., Herna´ndez, G. Essential role of MYB transcription factor: *PvPHR1* and microRNA: PvmiR399 in phosphorus-deficiency signaling in common bean roots. *Plant Cell Environ.* 2008; 31: 1834–1843.

Van Norman, J.M., Breakfield, N.W., Benfey, P.N. Intercellular communication during plant development. *Plant Cell* 2011; 23: 855–864.

Vatén, A., Dettmer, J., Wu, S., Stierhof, Y.D., Miyashima, S., Yadav, S.R., Roberts, C.J., Campilho, A., Bulone, V., Lichtenberger, R., Lehesranta, S., Mähönen, A.P., Kim, J.Y., Jokitalo, E., Sauer, N., Scheres, B., Nakajima, K., Carlsbecker, A., Gallagher, K.L., Helariutta, Y. Callose biosynthesis regulates symplastic trafficking during root development. *Dev. Cell* 2011; 21: 1144–1155.

Vidala, E.A., Arausa, V., Lub, C., Parryc, G., Greenb, P.J., Coruzzid, G.M., Gutiérreza R.A. Nitrate-responsive miR393/AFB3 regulatory module controls root system architecture in *Arabidopsis thaliana*. *PNAS*. 2010; 107: 4477–4482.

Voinnet, O. Origin, biogenesis, and activity of plant microRNAs. *Cell* 2009; 136:669–687.

Voinnet, O. Post-transcriptional RNA silencing in plant-microbe interactions: a touch of robustness and versatility. *Curr. Opin. Plant Biol.* 2008; 11: 464–470.

Wang, J.W., Wang, L.J., Mao, Y.B., Cai, W.J., Xue, H.W., Chena, X.Y. Control of root cap formation by microRNA-targeted auxin response factors in *Arabidopsis*. *Plant Cell* 2005; 17:2204–2216.

Weigel, D., Ahn, J.H., Blazquez, M.A., Borevitz, J.O., Christensen, S.K., Fankhauser, C., Ferrandiz, C., Kardailsky, I., Malancharuvil, E.J., Neff, M.M., Nguyen, J.T., Sato, S., Wang, Z.Y., Xia, Y., Dixon, R.A., Harrison, M.J., Lamb, C.J., Yanofsky, M.F., Chory, J. Activation tagging in *Arabidopsis*. *Plant Physiol.* 2000; 122: 1003–1013.

Wightman, B., Ha, I., Ruvkun, G. Posttranscriptional regulation of the heterochronic gene *lin-14* by *lin-4* mediates temporal pattern formation in *C. elegans*. *Cell* 1993; 75: 855– 862.

Williams, L., Carles, C.C., Osmont, K.S., Fletcher J.C. A database analysis method identifies an endogenous trans-acting short-interfering RNA that targets the *A. thaliana ARF2*, *ARF3*, and *ARF4* genes. *Proc. Natl. Acad. Sci. U. S. A.* 2005; 102: 9703–9708.

Windels, D., Vazquez, F. MiR393: integrator of environmental cues in auxin signaling? *Plant Signal. Behav.*. 2011; 6: 1672-1675.

Wu, G., Poethig, R.S. Temporal regulation of shoot development in *Arabidopsis thaliana* by miR156 and its target *SPL3*. *Development* 2006; 133: 3539–3547.

Wu, M.F., Tian, Q., Reed, J.W. *Arabidopsis* microRNA167 controls patterns of *ARF6* and *ARF8* expression, and regulates both female and male reproduction. *Development* 2006; 133:4211–4218.

Xie, F.L., Huang, S.Q., Guo, K., Zhu, Y.Y., Nie, L., Yang, Z.M. Computational identification of novel microRNAs and targets in *Brassica napus*. *FEBS Lett.* 2007; 581: 1464–1473.

Xie, Z., Allen, E., Fahlgren, N., Calamar, A., Givan, S.A., Carrington, J.C. Expression of *Arabidopsis* miRNA genes. *Plant Physiol.* 2005; 138: 2145–2154.

Xie, Z., Johansen, L.K., Gustafson, A.M., Kasschau, K.D., Lellis, A.D., Zilberman, D, Jacobsen, S.E., Carrington, J.C. Genetic and functional diversification of small RNA pathways in plants. *PLoS Biol.* 2004; 2: 642–652.

Xin, M., Wang, Y., Yao, Y., Xie, C., Peng, H., Ni, Z., Sun, Q. Diverse set of microRNAs are responsive to powdery mildew infection and heat stress in wheat (*Triticum aestivum* L.). *BMC Plant Biol.* 2010; 10: 123–134.

Yamaguchi, A., Wu, M.F., Yang, L., Wu, G., Scott Poethig R., Wagner, D. The microRNA regulated SBP-box transcription factor SPL3 is a direct upstream activator of *LEAFY*, *FRUITFULL*, and *APETALA*. *Dev. Cell.* 2009; 17: 268–278.

Yamasaki, H., Abdel-Ghany, S.E., Cohu, C.M. , Kobayashi, Y., Shikanai, T., Pilon, M. Regulation of copper homeostasis by microRNA in *Arabidopsis*. *J. Biol. Chem.* 2007; 282: 16369–16378.

Yamasaki, H., Hayashi, M., Fukazawa, M., Kobayashi, Y., Shikanai, T. SQUAMOSA promoter binding protein-like7 is a central regulator for copper homeostasis in *Arabidopsis*. *Plant Cell* 2009; 21: 347–361.

Yao, Y., Ni, Z., Peng, H., Sun, F., Xin, M., Sunkar, R., Zhu, J.K., Sun, Q. Non-coding small RNAs responsive to abiotic stress in wheat (*Triticum aestivum* L.). *Funct. Integr. Genomics* 2010; 10: 187–190.

Yoshikawa, M., Peragine, A., Park, M.Y., Poethig, R.S. A pathway for the biogenesis of trans-acting siRNAs in *Arabidopsis*. *Genes Dev.* 2005; 19: 2164–2175.

Yu, B., Yang, Z., Li, J., Minakhina, S., Yang, M., Padgett, R.W., Steward, R., Chen, X. Methylation as a crucial step in plant microRNA biogenesis. *Science* 2005; 307: 932–935.

Zhang, B.H., Pan, X.P., Cobb, G.P., Anderson, T.A. Plant microRNA: a small regulatory molecule with big impact. *Dev. Biol.* 2006; 289: 3–16.

Zhang, J., Xu, Y., Huan, Q., Chong, K. Deep sequencing of *Brachypodium* small RNAs at the global genome level identifies microRNAs involved in cold stress response. *BMC Genomics* 2009; 10: 449–465.

Zhao, B.T., Liang, R.Q., Ge, L.F., Li, W., Xiao, H.S., Lin, H.X., Ruan K.C., Jin, Y.X. Identification of drought-induced microRNAs in rice. *Biochem. Biophys. Res. Commun.* 2007; 354: 585–590.

Zhao, M., Ding, H., Zhu, J.K., Zhang, F., Li, W.X. Involvement of miR169 in the nitrogen-starvation responses in *Arabidopsis. New Phytol.* 2011; 190: 906–915.

Zhou, L., Liu, Y., Liu, Z., Kong, D., Duan, M., Luo, L. Genome-wide identification and analysis of drought-responsive microRNAs in *Oryza sativa. J. Exp. Bot.* 2010; 61: 4157–4168.

Zhou, X., Wang, G., Sutoh, K., Zhu, J., Zhang, W. Identification of cold-inducible microRNAs in plants by transcriptome analysis. *BBA-Gene Regul. Mech.* 2008; 1779: 780–788.

Zhou, Z.S., Song, J.B., Yang, Z.M. Genome-wide identification of *Brassica napus* microRNAs and their targets in response to cadmium. *J. Exp. Bot.* 2012b; 63: 4597–4613.

Zhou, Z.S., Zeng, H.Q., Liu, Z.P., Yang, Z.M. Genome-wide identification of *Medicago truncatula* microRNAs and their targets reveals their differential regulation by heavy metal. *Plant Cell Environ.* 2012a; 35: 86–99.

In: MicroRNA and Non-Coding RNA
Editor: James C. Johnson
ISBN: 978-1-62618-442-8

Chapter 2

MicroRNAs, Major Affective Disorders and Suicidal Behaviour

Gianluca Serafini[1*], Maurizio Pompili[1], Marco Innamorati[1], Yogesh Dwivedi[2], Noam Shomron[3] and Paolo Girardi[1]

[1]Department of Neurosciences, Mental Health and Sensory Organs, Suicide Prevention Center, Sant'Andrea Hospital, Rome, Italy

[2]Psychiatric Institute, Department of Psychiatry, College of Medicine, University of Illinois at Chicago, Chicago, IL, US

[3]Department of Cell and Developmental Biology and Functional Genomics Laboratory, Sackler Faculty of Medicine, Tel Aviv University, Tel Aviv, Israel

Abstract

Major affective disorders are common and disabling conditions associated with significant psychosocial impairment and suicidal risk in the general population. At least 3–4% of all depressive individuals die by suicide. At a molecular level, affective disorders and suicidal behaviour have been recently associated with abnormalities in structural and synaptic plasticity. A recent hypothesis suggested that small non-coding RNAs (ncRNAs), in particular microRNAs (miRNAs), play a critical role in the translational regulation at the synapse. Here we summarized the current literature on miRNAs putative subcellular localization and sites of action in mature neurons analyzing their role in neurogenesis, synaptic plasticity, pathological stress changes, major affective disorders and suicidal behaviour. miRNAs have played a fundamental role in the evolution of brain functions. The perturbation of some intracellular mechanisms as well as impaired assembly, localization, and translational regulation of specific RNA binding proteins may affect learning and memory, presumably contributing to the pathogenesis of major affective disorders and perhaps suicidal behaviour. Also, miRNA dys-regulation has also been associated with several neuropsychiatric diseases. However, further

* Correspondence: Gianluca Serafini M.D., Ph.D., Department of Neurosciences, Mental Health and Sensory Organs, Sant'Andrea Hospital, Sapienza University of Rome, 1035-1039 Via di Grottarossa, 00189, Rome, Italy. Tel: +390633775675; Fax: +390633775342. E-mail: gianluca.serafini@uniroma1.it.

evidence are needed in order to directly clarify the role of miRNAs in major affective disorders and suicidal behaviour [Abstract words (189); keywords (4): microRNAs; synaptic plasticity; major affective disorders; suicidal behaviour].

Introduction

Major depression and suicide are relevant public health concerns. Major Depressive Disorder (MDD) is a common and disabling psychiatric condition affecting approximately 2.5% of the general population. MDD has been suggested to become the second highest burden of disease (measured in disability-adjusted life years) by 2020 (World Health Organization, 2012a). MDD has negative social consequences in terms of reduced employment and psychosocial impairment (Andersen et al, 2011). Also, suicidal behaviour is a major cause of death and morbidity worldwide and it is globally among the top 20 leading causes of death for all ages. Approximately 1 million people died by suicide every year and for every completed suicide there are at least between 10 and 40 attempted suicides in the world (World Health Organization, 2012b). Despite many years of research, the molecular and cellular mechanisms associated with both major depression and suicidal behaviour remain quite unclear. Recent studies suggest that alterations in the expression of genes, including those which regulate neural and structural plasticity, may be crucial in the pathogenesis of major affective disorders (Dwivedi, 2009; Serafini et al, 2011, 2012).

During the last years, research aimed to investigate the complex relationship between genes and proteins expression, particularly the correlation between genes encoding messenger RNAs (mRNAs) and neuropsychiatric disorders. miRNAs are a relevant class of gene expression regulators involved in the development, physiology, and diseases of the central nervous system (Bushati and Cohen, 2007; Fiore et al, 2008; Mouillet-Richard et al, 2012). Throughout the evolution miRNAs served as metacontrollers of gene expression responsible for cellular changes fundamental for brain development and cognitive functioning.

Understanding the molecular mechanisms regulating miRNAs expression is crucial in order to investigate the pathophysiology of neuropsychiatric disorders (Dwivedi, 2011; Saugstad, 2010). An activity-regulated transcriptional program in neurons consisting of hundreds of genes is critical for the development of neural circuits, activity-dependent changes and neural connectivity as reported by Flavell and Greenberg (2008). The authors found that genes encoding retrograde signals to the presynaptic cell (such as brain-derived neurotrophic factor (BDNF)) or truncated, dominant-interfering forms of full-length proteins at the synapse level (such as homer1a), strongly influence dendritic and synaptic remodeling. Hypothesizing that miRNAs modulate a large variety of gene expression patterns during development and tissue homeostasis, Newman and Hammond (2010) investigated the specificity of the Microprocessor (the protein complex essential for maturation of canonical miRNAs) and the miRNA families requiring an interaction between RNA-binding proteins and cis-regulatory sequences within miRNA precursor loops. Post-transcriptional control of miRNA expression may be only partially regulated by the miRNA biogenesis pathway and largely depends on the interactions between the pri-miRNA and coactivator proteins providing a critical specificity.

miRNA altered regulation may impair neurophysiological functions like learning and memory that have been recognized to play a crucial role in major depression. Being master regulators of gene expression, miRNAs appear to be relevant genetic easily accessible biomarkers for the diagnosis, treatment and progression of human diseases.

Considerable evidence suggest that miRNAs may be implicated in neuropsychiatric disorders, such as Huntington disease (Johnson et al, 2008), Parkinson disease (Kim et al, 2007), and Tourette's syndrome (Abelson et al, 2005). Recently, several studies have reported abnormal levels of specific miRNAs in postmortem brain of subjects with a diagnosis of schizophrenia (Beveridge et al, 2008, 2010; Perkins et al, 2007; Kim et al, 2010). Interestingly, these alterations seem to be non-dependent by antipsychotic treatment (Perkins et al, 2007).

Also, single nucleotide polymorphic analysis identified two brain-expressed miRNAs in a case-control study as being linked to schizophrenia (Hansen et al, 2007) and it has been clearly demonstrated that stress, glucocorticoids and mood-stabilizers modulate expression of selective miRNAs (Hunsberger et al, 2009).

Measurements of miRNAs may substantially contribute to our understanding of how gene expression networks are reorganized in major depression and/or suicide. However, whether the miRNA changes may lead to altered expression of prefrontal cortex mRNAs, directly (by acting on miRNA targets) or indirectly (e.g., by affecting a gene that acts on a miRNA target, such as a transcription factor) is still quite unclear. There are also, to our best knowledge, no other direct studies investigating the role of brain miRNAs in major depression and suicide.

Based on the current literature, the present review chapter is aimed to summarize the most relevant biochemical information about the role of miRNAs in neurogenesis and neuroplasticity mechanisms and their possible involvement in the development of pathological stress changes, affective disorders and suicidal behaviour.

2. Methods

A reference search was performed across the Medline and ScienceDirect databases (January 1980 and December 2011). The search used the following terms: "microRNAs" AND "Neuroplasticity" OR "Synaptic plasticity" AND "Neurogenesis" AND "Affective Disorders" AND "Major Depression" OR "Major Depressive Disorder (MDD)" AND "Suicid*" (including suicidal behaviour OR suicide ideation OR suicidal thoughts OR deliberate self harm OR suicidal attempt). In addition, the reference lists of all papers identified were reviewed. Included papers were those published in peer reviewed journals.

Where a title or abstract seemed to describe a study eligible for inclusion, the full article was obtained and examined to assess its relevance. Two independent researchers conducted a two-step literature search. Any discrepancies between the two reviewers who, blind to each other, examined the studies for the possible inclusion were resolved by consultations with a senior author.

Biogenesis and Expression of miRNAs, Mechanisms of miRNA-Mediated Translational Inhibition and Possible Sites of Dysfunction

In the brain, gene expression may be regulated at multiple levels. Gene expression may be activated through transcriptional factors and alternative splicing but also by several non-coding RNA transcripts creating miRNAs, antisense RNAs, and other critical forms of RNAs (for more details see Dwivedi, 2011;Serafini et al, 2012). miRNAs may act as regulators of developmental timing and cell fate (Kosik, 2006; Bushati and Cohen, 2007). Approximately 2000 miRNAs have been identified in the human genome (miRBase, 2012); they are usually expressed in tissue and cell specific manner (Landgraf et al, 2007; Kapsimali et al, 2007; Choi et al, 2008). Specifically, the last version of miRBase database includes 21264 hairpin precursor miRNAs entries expressing 25141 mature miRNA in 193 species (Kozomara and Griffiths-Jones, 2011).

Approximately 20–40% miRNAs are expressed together during the brain development (Miska et al, 2004; Sempere et al, 2004).

miRNAs are evolutionarily conserved as implicated in various processes including the stress response (He et al, 2007). Smirnova et al. (2005) hypothesized that some miRNAs are expressed in neurons, whereas others in high concentrations in astrocytes (Mor et al, 2011). Several evidence (Presutti et al, 2006; Fiore et al, 2008; Liu and Zhao, 2009; Vreugdenhil and Berezikov, 2009) suggested that miRNAs may locally regulate mRNA translation inducing changes in neurogenesis, synaptic and axon development, neuronal plasticity. Also, miRNAs may contribute to the onset and maintenance of affective disorders at multiple levels as a result of abnormalities in synaptic plasticity and neurogenesis or by regulating the expression of genes critically involved in major depression (Dwivedi, 2011).

miRNAs are encoded in primary miRNAs (pri-miRNA) and first processed by mostly RNA polymerase II but also polymerase III and then by RNase III enzyme Drosha within the nucleus to create miRNA precursors (pre-miRNAs). Pre-miRNAs are also processed in the cytoplasm by the RNase III enzyme Dicer requiring cofactors such as human transactivation response (TAR), RNA-binding protein (TRBP) or RNA-activated protein kinase (PKR)-activating protein (PACT). RNA-induced silencing complex binds to specific sequences within the 3'-UTR region of mRNAs and can interfere with translation and levels of the mRNA. The inhibition of translation may be mediated by AGO (having a splicer function). Another mechanism is that RNA-induced silencing complex stored target mRNA through the action of exonucleases, decapase, and deadenylase in a translationally inactive state. Also, miRNA may regulate gene transcription through a downregulation of transcription factors leading to reduced protein synthesis (Michalak, 2006).

The presence of a specific miRNA does not automatically ensure that it is functional (for more details, see Figure 1 in McNeill and Van Vactor, 2012). This depends on the threshold of the miRNA level, its cellular location or compartment, the presence of a target mRNA, the stoichiometric ratio between the miRNA and mRNA target, RNA co-factors binding to the target mRNA, and possibly cellular activation events such as raised calcium levels or proteolytic cleavage or phosphorylation of miRNA pathway protein components (McNeill and Van Vactor, 2012; Figure 2 in Serafini et al, 2012). Potential reasons able to account altered miRNA expression are genetic changes in the promoter region of pri-miRNA gene transcripts, the pre-miRNA secondary hairpin structure, or the mature miRNA structure (Michalak, 2006). RNA editing (Alon et al, 2012) of transcripts or epigenetic suppression of the chromosomal region encoding the miRNAs might also play a role in the miRNA control. Finally, changes in the levels of the miRNA machinery (e.g., Dicer, Drosha, or RISC) also account for miRNA levels (Michalak, 2006).

miRNAs, Neurogenesis and Neuroplasticity

Environmental stimuli, exercise, and antidepressant drugs may enhance hippocampal neurogenesis (Lazarov et al, 2010) that is involved in learning and memory (Bekinschtein et al, 2011; Koehl and Abrous, 2011; Yirmiya and Goshen, 2011) whereas pathological stress and major depression negatively affect hippocampal neurogenesis (DeCarolis and Eisch, 2010; Lucassen et al, 2010). Recent evidence hypothesized an active role of miRNAs in the regulation of synaptic plasticity, neurogenesis, and neuropsychiatric disorders.

miR-124, for example, has been reported as a fundamental regulator of the temporal progression of adult neurogenesis (Cheng et al, 2009). The progression along the subventricular zone stem cell lineage to neurons is mediated by the regulation of the sex determining region (Sox9), induced by miR-124. Overexpression of Sox9 blocked neuronal differentiation whereas Sox9 knockdown enhanced neuronal formation. More than 100 specific miRNAs such as miR-124a and let-7, miR-200 variants seem to be expressed by olfactory tissues (Choi et al, 2008).

miR-137, which is regulated by Sox2 has been suggested to be critical for the proliferation and differentiation of adult neural stem cells (Szulwach et al, 2010). MeCP2 (a DNA methyl-CpG-binding protein) epigenetically induced the overexpression and reduction of miR-137 that enhanced both the proliferation and differentiation of adult neural stem cells. The expression of Ezh2 (a histone methyltransferase and Polycomb group protein) has been suggested to be post-transcriptionally inhibited by miR-137 with the resulting reduction in histone H3 trimethyl lysine 27 (Szulwach et al, 2010). The existence of a possible association between miRNAs and epigenetic mechanisms regulating adult neurogenesis has been demonstrated by the link between Ezh2 and miR-137. Many more mechanisms are expected to exist.

A critical role in regulating synaptic plasticity through the action on specific targets has been suggested for miRNAs (Kosik, 2006; Gao, 2008; Christensen and Schratt, 2009). Malan-Müller et al. (2012) reported that the role of miRNAs as translational regulators suggests these small RNAs as mediators of long-term plasticity. miRNA regulation of dendritic spine development in rats has been analyzed by Schratt et al. (2006).They suggested that most dendritic spines are implicated in postsynaptic signaling/plasticity and include a postsynaptic density (Javorski et al, 2009).

Behavioural, learning, dendritic spine, and long-term potentiation (LTP) deficits were affected by the loss of the fragile X mental retardation protein (FMRP) which regulates protein synthesis in dendritic spines (Jin et al, 2004) and interacts with RNA-induced silencing complex (RISC) components, Argonaute-1 (Ago1) and the RNA endonuclease Dicer. Different components of miRNAs, such as pre-miRNAs, Dicer, and eukaryotic translation initiation factor-6 (eIF2c) protein may modulate protein synthesis by binding to specific mRNAs sites presumably through RISC in the somatodendritic compartment (Qurashi et al, 2007; Caudy et al, 2002; Ishizuka et al, 2002). Also, RISC down-regulated synaptic expression of calcium/calmodulin-dependent protein kinase IIa (CaMKIIa), having binding sites for miR-280 and miR-289 (Ashraf et al, 2006). The same authors suggested that Armitage plays a critical rolefor LTP and synaptic protein synthesis. Similarly, the proteasomal degradation of MOV10, an RNA helicase associated with Argonaute protein, was reported to play a relevant role in synaptic plasticity regulation (Banerjee et al, 2009).

miR-134 has been reported to regulate rat protein synthesis together with the size of postsynaptic sites in dendritic spines of the synaptodendrites located in hippocampal neurons (Schratt et al, 2006). miR-134 suppression which is mediated by the translation of mRNA encoding a specific protein kinase, Limdomain-containing protein kinase 1 (Limk1) regulates dendritic spine growth whereas miR-138 down-regulates the size of dendritic spines in rat hippocampal neurons affecting the expression of acyl protein thioesterase 1 involved in the palmitoylation status of proteins having synaptic functions within dendrites (Siegel et al, 2009). The suppression of miR-138 induced spine growth which may be inhibited by interference-mediated knockdown of acyl protein thioesterase 1 and expression of membrane-localized Ga13.

Considering that mature miRNAs are partially processed near synapses, their fundamental role in local protein synthesis was suggested (Lugli et al, 2008). miRNAs derive at least partially by processing miRNA precursors within dendritic spines. Near the synapse, cAMP response element-binding (CREB) responsible for long-term changes of gene expression regulates a local processing of miRNA precursors. miRNA genes like miR-132 have binding sites in their promoter regions for Creb1 and their activation is induced by BDNF through CREB (Vo et al, 2005;Wu and Xie, 2006). Neuronal activity has been reported to regulate spine formation (Impey et al, 2010), at least partially by increasing miR-132 transcription. Neurite outgrowth/dendritic morphogenesis may be enhanced by miR-132 overexpression (Vo et al, 2005; Wayman et al, 2008) and increases miR-124 having a critical role in neuronal cell identity. Interestingly, CREB may regulate miR-124 during Dicer processing, RISC incorporation and stabilization by Argonaute proteins through the action of 5-HT locally at synapses (Rajasethupathy et al, 2009). miR-124 has been regulated by 5-HT de-repressing CREB and enhancing 5-HT-dependent long-term facilitation.

Furthermore, SIRT1, a gene playing a fundamental role in oxidative stress and circadian rhythms (Nakahata et al, 2008; Finkel et al, 2009), has been suggested to regulate synaptic plasticity and memory formation in miR-134 acting on post-transcriptional CREB expression (Gao et al, 2010). An unchecked miR-134 expression and consequent reduced CREB and BDNF expressions determining an impaired synaptic plasticity may be induced by SIRT1 reduction.

A progressive loss of brain specific miRNAs was reported in differentiated adult neurons of Dicer 1-mutant mice (in which a Dicer1 gene deletion has been induced) as well as an enhancement of cognition (presumably mediated by increased translation of proteins, such as BDNF and MMP-9) having a crucial role in synaptic morphology (Konopka et al, 2010). Other evidence suggested that critical phases of memory formation are affected by miRNAs via synaptic tagging aimed to establish synaptic input specificity (Kim et al, 2004; Martin and Kosik, 2002; Schaeffer et al, 2003).

Very recently, Malan-Müller et al. (2012) suggested that the critical role of miRNAs in neurogenesis, neurite outgrowth, synaptogenesis and synaptic and neural plasticity together with their regulatory role under stressful conditions strongly support their relevance in the pathophysiology of anxiety disorders. Aiming to examine the role of miR-206 in regulating BDNF and memory function in Alzheimer disease mice, Lee et al. (2012) demonstrated increased levels of miRNAs in the brains of Tg2576 mice and the temporal cortex of human heimer disease brains. They found that AM206, a neutralizing inhibitor of miR-206 (antagomir), increased brain levels of BDNF into the mice brains improving memory function and preventing the detrimental effects of amyloid-β42. Also, AM206 hippocampal synaptic density and neurogenesis were enhanced by AM206. Bocchio-Chiavetto et al. (2012) suggested that thirty miRNAs were differentially expressed after 12 weeks of treatment with escitalopram using the entire miRNA repertiore for the analysis (qRT-PCR) of the changes in the blood of 10 depressed subjects. The authors found that there was a significant enrichment in neuroactive ligand-receptor interaction, axon guidance, long-term potentiation and depression confirming that miRNAs may be differentially involved in the regulation of antidepressant mechanism.

Also, Pathania et al. (2012) suggested that miR-132 represents the basis of a structural plasticity observed in postnatal neurogenesis and miR-132 overexpression in transplanted neurons may enhance neuronal survival. The authors reported a significant increase in the survival of newborn neurons, specifically a timely directed overexpression of miR-132 at the onset of synaptic integration using in vivo electroporation. In summary, based on the many studies mentioned above, we believe it is substantiated that miRNAs play a critical role in neurogenesis and neuroplasticity mechanisms.

miRNAs, BDNF and Major Affective Disorders

Neuronal differentiation and cell survival are known to be promoted during development by BDNF, a member of the neurotrophin family of structurally related proteins (Cowansage et al, 2010). In particular, cell survival, migration, outgrowth of axons and dendrites, synaptogenesis, and remodeling of synapses are promoted by BDNF-TrkB signaling pathway (Ohira and Hayashi, 2009). Several aspects of synaptic plasticity such as morphological plasticity, neuronal proliferation, migration, and phenotypic differentiation are regulated by BDNF (McAllister et al, 1999; McAllister, 2001; Thoenen, 2000) which is involved in the maintenance of various neuronal functions and neurogenesis (Cooper et al, 1996; Sofroniew et al, 1990). Changes in BDNF expression, release, and activity may be regulated by epigenetic and post-translational mechanisms associated with structural brain abnormalities, reduced neural plasticity and neuropsychiatric conditions. miR-132 has been suggested to regulate the X-linked transcriptional repressor methyl CpG-binding protein 2 (MeCP2) translation (Klein et al, 2007).

Changes in MeCP2 levels have been found in several neurodevelopmental defects like Rett syndrome. MeCP2 and BDNF levels in cultured rat neurons were increased by the blocking of miR-132-mediated repression while BDNF and miR-132 levels in vivo were reduced by the loss of MeCP2. MeCP2 has been suggested as a critical regulator of neuroplasticity in postmitotic neurons (Im et al, 2010). Based on recent evidence (Im et al, 2010), MeCP2 controlled the effects of cocaine on dorsal striatal BDNF levels by interacting with miR-212. Therefore, homeostatic interaction between MeCP2 and miR-212 could regulate vulnerability to cocaine addiction.

Several researchers (Smith et al, 1995; Dwivedi et al, 2003b; Karege et al, 2005; Brunoni et al, 2008; Castrén and Rantamäki, 2008; Kozisek et al, 2008; Dwivedi et al, 2009a, 2009b; Fanous et al, 2010; Shi et al., 2010) suggested that reduced BDNF levels may be associated with severity of depressive symptoms in major depression. Notably, neurotrophin signaling promoting neuronal and synaptic remodeling as well as neurogenesis in the hippocampus and prefrontal cortex may be increased by some antidepressant drugs (Duman, 2004; Castrén et al, 2007; Sairanen et al, 2007; Bessa et al, 2009; McEwen and Chattarji, 2004).

BDNF-induced synaptic maturation and excitatory neurotransmitter glutamate release have been recently found to be inhibited by glucocorticoids (Kumamaru et al, 2008; Numakawa et al, 2009) whereas BDNF and glutamate release may be enhanced by antidepressant medications (Yagasaki et al, 2006). The expression of glucocorticoid receptors (GRs) was reported to be down-regulated in depressed subjects (Pariante and Miller, 2001) especially in the paraventricular nucleus (PVN) of F344 rats (Uchida et al, 2008). Conversely, a specific role of miR-18a in regulating GR protein expression has been hypothesized considering that the expression of miR-18a, inhibiting the translation of GR mRNA, is found to be greater in those rats. Furthermore, miR-18a and miR-124a have been found to control GRs expression inducing a down-regulation and translation of GRs, implicated in stress-related disorders (Vreugdenhil et al, 2009). In particular, GR levels were reported to reduce the expression of miR-124a and miR-18a through the induction of the gene leucine zipper. Possible miRNA binding sites regulating GR levels may be included in the GR first exon (Turner et al, 2010).

However, the real effect of miR-124a expression on the GR levels is doubtful as an effect only in the case of miR-124a but not for miR-18 overexpression was observed by Vreugdenhil et al. (2009) after conducting the luciferase assays using the GR 3'-UTR. Several authors (David et al, 2009; Huang and Herbert, 2006) suggested that glucocorticoids may be involved in the neurogenic action of antidepressants. Recently, Anacker et al. (2011) reported that AD-induced changes in neurogenesis are directly dependent on the GRs. In addition, BDNF has been found to regulate the expression of miR-132 implicated in neurite outgrowth and dendritic morphogenesis through the action of CREB (Kawashima et al, 2010). The ERK1/2 pathway activation determines the up-regulation of miR-132 induced by BDNF levels. Both BDNF and ERK 1/2 are implicated in pathological stress-induced changes and major depression. miRNAS such as miR-30a and miR-195 correlate with BDNF levels (Mellios et al, 2011) which are associated with major depression. Huang et al. (2012) suggested that BDNF rapidly elevated Dicer, increasing mature miRNA levels and inducing RNA processing bodies in neurons and rapidly induced Lin28, causing selective loss of Lin28-regulated miRNAs and a corresponding upregulation in translation of their target mRNAs. Additionally, BDNF translation specificity and BDNF-dependent dendrite arborization were inhibited by Lin28 deficiency, or Lin28-resistant Let-7 precursor miRNA. The authors concluded that the specificity in BDNF-regulated translation is linked to two-part posttranscriptional control of miRNA biogenesis enhancing mRNA repression together with GW182 (selectively derepressing and increasing translation of specific mRNAs). Therefore, miRNAs may exert a crucial role in the pathophysiology of major depression.

miRNAs and Stress-Induced Changes

Acute stress may rapidly influence miRNA expression and synaptic transmission in frontal cortex regulating mRNA translation. As recently mentioned (Kagias et al, 2012), environmental stress has an impact on neuronal differentiation and maintenance at both the molecular and cellular levels. Changes in the expression of several molecules such as transcription factors and miRNAs have been suggested to regulate stress resistance and adaptation.

Rinaldi et al. (2010) reported that, after acute stress, the expression of several miRNAs was increased in the frontal cortex but not in the hippocampus of mice. In particular, after repeated stress the expression of let-7a, miR-9, and miR-26a/b resulted significantly increased. Chronic stress determines larger changes in the expression of miR-134, miR-183, miR-132, Let-7a-1, miR-9-1, and miR-124a-1 in the central amygdala and hippocampal CA1 region compared to acute stress (Meerson et al, 2010). Under acute or chronic stress, miR-376b and miR-208 levels resulted increased whereas miR-9 was decreased in the hippocampal CA1 region. Conversely, both acute and chronic stress modulated the expression of miR-Let-7a-1 in the central amygdala. miR-134 and miR-183 target various splicing factors with the final result of a modulation of alternative splicing. The splicing factor SC35, able to promote the alternative splicing of acetylcholinesterase, was increased by knockdown of miR-183 expression and decreased by overexpression of miR-183. miR-183 also targets profilin-2 mRNA (Witke, 2004), the expression of which has been suggested to modify neurotransmitter homeostasis and behaviour presumably regulating dendritic spine morphology in neuronal cells.

The repressor element-1 silencing transcription factor (REST4) is a critical element in the development of stress-related disorders in adulthood. The expression of REST4 determining depression-like behaviours was increased by maternal separation (both mRNA and proteins levels) in the medial prefrontal cortex of neonatal mice (Uchida et al, 2010). Furthermore, REST may regulate the expression of several brain-enriched miRNAs (Conaco et al, 2006; Otto et al, 2007) implicated in the brain development and plasticity (Vo et al, 2005; Kosik, 2006; Rajasethupathy et al, 2009). REST has been suggested to control in the medial prefrontal cortex of stressed rats the expression of pre-miRNAs-132 and 124 with a predominant role in synaptic plasticity (Vo et al, 2005; Rajasethupathy et al, 2009), to reduce dendritic length (Pascual et al, 2007), and to increase synaptic density (Ovtscharoff and Braun, 2001) together with interfering with basal neuronal activity (Stevenson et al, 2008).

Recently, Soldati et al. (2012) found that the aberrant repression of several miRNAs, including miR-137, is directly mediated by REST, the association of which with Huntingtin in the cytoplasm is found to be disrupted in Huntington's disease, with the final result of increased nuclear REST and repression of several neuronal-specific genes, including BDNF. Also, Nerini-Molteni et al. (2012) investigating miRNA expression profiles in a human pluripotent cell model found an association between methylmercury chloride, a well known brain developmental toxicant, and the ubiquitin-proteasome protein degradation pathway. They reported significant changes in the expression of 12 miRNAs, several predicted targets for the observed miRNAs and functional clusters associated with neural development. Also, they suggested that apoptosis or stress response-related genes were indicative of general toxic effects. The authors concluded that miRNA expression analysis may be considered a powerful instrument in pathway-oriented toxicity.

In overall, homeostatic miRNA response after stress has been reported to modulate the expression of transcriptional factors such as CREB1. Another observation must be mentioned. Desipramine and fluoxetine may reverse learned helplessness (LH) behaviour (Dwivedi et al, 2004, 2005b, 2005c) and LH decrease of hippocampal neurogenesis (Chen et al, 2006) as suggested by several animal studies including repeated shock sessions and able to prolong the duration of LH. Lugli et al. (2008) reported adaptive transcriptionally controlled downregulations of miRNA in response to inescapable shock in non-LH rats and a significant blunted response in LH rats (Dwivedi et al, 2004, 2005b, 2005c).

A possible homeostatic response aimed to minimize repressive effects on Creb1 has been suggested as miR-200a, miR-200b, miR-200c targeting Creb1 resulted down-regulated only in non-LH rats. It has been hypothesized that normal homeostatic miRNA response to repeated inescapable shock is not blunted in LH rats but actively reorganized (independently of stress response) through gene expression networks.

The Role of miRNAs in Major Depression and Suicidal Behaviour

Some studies suggested the existence of miRNAs dysfunctions in patients with major depression. Some miRNAs, regulating brain-active transcription factors such as NOVA1, signaling proteins, and genes implicated in neurotransmitter release and synaptic plasticity, were reported to be significantly down-regulated in the prefrontal cortex of subjects with major depression (Dwivedi and Smalheiser, unpublished findings). The expression of VEGFA, a specific growth factor involved in both animal and human depression, together with other critical signaling proteins, ion channels, and ubiquitin ligases, seem to be down-regulated by various miRNAs (among cell cycle-related genes). Also, Xu et al. (2010) investigated the association between miRNA polymorphisms and vulnerability to major depression. P300 latency induced by an auditory oddball task resulted significantly associated with miR-30e ss178077483 genotypes (C/T genotype) and with the frequency of miR-30e ss178077483 in depressed subjects in a sample of 1088 Chinese depressed patients and 1102 control subjects.

However, miRNAs dysfunctions were also found in patients with bipolar disorder (BD). The expression of fifteen miRNAs are both up- and down-regulated in the prefrontal cortex of subjects with BD predicting the involvement of approximately 2666 genes in neuropsychiatric functions related to the disease (Kim et al, 2010). A significantly reduced expression of miR-134 after acute treatment as well as its increased plasmatic levels after 2 and 4 weeks of treatment were reported in the peripheral cells of 21 drug-free patients with BD type I compared to 21 matched healthy controls (Rong et al, 2011). Similar findings were replicated in animal models receiving psychoactive treatments. miRNAs and their effectors are also important targets for the action of psychoactive drugs (Zhou et al, 2009).

Hippocampal miRNA levels have been suggested to be altered after chronic lithium and valproate treatment. The predicted effectors of these miRNAs have been reported to regulate neurite outgrowth, neurogenesis, and important signaling pathways. Levels of miR-34a and increased levels of the predicted effector metabotropic glutamate receptor-7 were reduced after treatment with lithium or valproate. Also, metabotropic glutamate receptor 7 levels were reduced by miR-34a precursor and increased by a miR-34a inhibitor. Machado-Vieira et al. (2010) suggested that signaling pathways such as protein kinase C (PKC), phosphatase and tensin homolog (PTEN), extracellular signal-regulated kinase mitogen-activated protein (ERK-MAP kinase), Wnt/b-catenin (playing an important role in both embryonic development and tumorigenesis), and adrenergic signaling pathways are the identified miRNAs targets involved in both major depression and BD. Notably, dipeptidylpeptidase- 10 (DPP10), glutamate receptor-7 (Saus et al, 2010) and thyroid hormone receptor beta (THRB) hippocampal protein levels were significantly up-regulated whereas the expression of miR-128a, miR-24, and miR-34a was down-regulated after chronic administration of lithium and valproate. Kuss and Chen (2008) suggested miRNA-mediated effects at the synapse with a direct influence on transcripts through a stimulation-dependent translation, similarly to ionotropic glutamate receptors stimulated by glutamate (Kuss and Chen, 2008; Kosik, 2006).

Other researchers examined the effect of lithium on miRNAs. Chen et al. (2009) reported that, in 20 lymphoblastoid cell lines in culture, the expression of miR-221, miR-152, miR-15a, miR-494, miR-155, miR-181c, and miR-34a was persistently altered after more than 15 days of lithium treatment. Also, the effects of some ADs seem to be mediated by modifications of specificmiRNAs. Baudry et al. (2010) suggested that miR-16 may play a critical role in regulating 5-HT transporter expression as well as in mediating adaptive responses of serotonergic and noradrenergic neurons to fluoxetine treatment. An inverse correlation was suggested between miR-16 levels, highly expressed in noradrenergic neurons, and 5-HT transporter expression. After long-term treatment with fluoxetine (Baudry et al, 2010), increased miR-16 levels and reduced 5-HT transporter expression may be observed in serotonergic raphe nuclei of mice. A release of S100b reducing miR-16 levels and an unlocked expression of serotonergic functions in the locus coeruleus of noradrenergic neurons have been found after long-term fluoxetine. Both ADs and mood stabilizers seem to have an active role on miRNA levels. The translation of Bcl-2 has been suggested to have anti-apoptotic functions and to play a critical role in affective disorders (Cimmino et al, 2005) and miR-15 and miR-16 seem to be inhibited by mood stabilizers (Marmol, 2008; Kim et al, 2010).

ADs and mood stabilizers may use miRNAs as downstream effectors (O'Connor et al, 2012). Based on preclinical models, modified behaviours were associated with modifying miRNA levels. Recently, Millan (2011) confirmed the central role of miRNAs as important effectors in regulating serotonergic transmission (having important implications in the development of major depression). Dwivedi et al. (2003a) have previously reported that miR-124, whose expression was modulated by 5HT (Rajasethupathy et al, 2009), regulates the expression of CREB (having a CREB site within its promoter region), a crucial transcription factor involved in major depression. The A-element able to confer repression by miR-96 has been reported within mRNA of 5HT1B receptor by Jensen et al. (2009). A common human variant (G-element) reduces the repressive activity of this element by disrupting a nucleotide revealing critical for its interaction with miR-96. Jensen et al. (2009) reported that in a sample of 359 college students, individuals who were homozygous for the ancestral A-element have more conduct disorder behaviours than individuals with the G-element.

Table 1. Candidate genes, predicted miRNAs and prediction software involved in suicide

Predicted miRNAs	Candidate genes involved in suicide	Author(s), year	Prediction software
miRNA Has-miR-185	Tyrosine receptor kinase type B-T1, a truncated TrkB transcript	Maussion et al., 2012	The RNA22 (http://cbcsrv.watson.ibm.com/rna22_targets.html) database was used to characterize differential expressed miRNAs that could bind Trkb-T1 3'UTR sequence (for more details see Miranda et al., 2006)).
has-miR-137	CDH6, E2F6, NCOA2	Smalheiser et al., 2012	Combination of evidence from 11 miRNA target prediction servers using the miRecords meta-server. For each downloaded miRNA, the meta-server was queried to identify targets that are predicted by at least 3 different prediction servers, and the top 30 ranked targets were tabulated (a small fraction of the total number of predictions).
has-miR-148b	DNMT3B	Smalheiser et al., 2012	Combination of evidence from 11 miRNA target prediction servers using the miRecords meta-server. For each downloaded miRNA, the meta-server was queried to identify targets that are predicted by at least 3 different prediction servers, and the top 30 ranked targets were tabulated (a small fraction of the total number of predictions).
has-miR-101	EZH2, MYCN, ICOS	Smalheiser et al., 2012	Combination of evidence from 11 miRNA target prediction servers using the miRecords meta-server. For each downloaded miRNA, the meta-server was queried to identify targets that are predicted by at least 3 different prediction servers, and the top 30 ranked targets were tabulated (a small fraction of the total number of predictions).
has-miR-335	SOX4, PTPRN2, MERTK	Smalheiser et al., 2012	Combination of evidence from 11 miRNA target prediction servers using the miRecords meta-server. For each downloaded miRNA, the meta-server was queried to identify targets that are predicted by at least 3 different prediction servers, and the top 30 ranked targets were tabulated (a small fraction of the total number of predictions).
has-miR-376a	SLC16A1, SFRS11, TTK	Smalheiser et al., 2012	Combination of evidence from 11 miRNA target prediction servers using the miRecords meta-server. For each downloaded miRNA, the meta-server was queried to identify targets that are predicted by at least 3 different prediction servers, and the top 30 ranked targets were tabulated (a small fraction of the total number of predictions).
has-miR-155	AGTR1, BACK1, LDOC1, MATR3, TM6SF1	Smalheiser et al., 2012	Combination of evidence from 11 miRNA target prediction servers using the miRecords meta-server. For each downloaded miRNA, the meta-server was queried to identify targets that are predicted by at least 3 different prediction servers, and the top 30 ranked targets were tabulated (a small fraction of the total number of predictions).

Predicted miRNAs	Candidate genes involved in suicide	Author(s), year	Prediction software
has-miR-130a	TAC1, CSF1, MAFB, MEOX2, HOXA5	Smalheiser et al., 2012	Combination of evidence from 11 miRNA target prediction servers using the miRecords meta-server. For each downloaded miRNA, the meta-server was queried to identify targets that are predicted by at least 3 different prediction servers, and the top 30 ranked targets were tabulated (a small fraction of the total number of predictions).
has-miR-27a	SP1, SP3, SP4	Smalheiser et al., 2012	Combination of evidence from 11 miRNA target prediction servers using the miRecords meta-server. For each dowloaded miRNA, the meta-server was queried to identify targets that are predicted by at least 3 different prediction servers, and the top 30 ranked targets were tabulated (a small fraction of the total number of predictions).
has-miR-10a	HOXA1	Smalheiser et al., 2012	Combination of evidence from 11 miRNA target prediction servers using the miRecords meta-server. For each downloaded miRNA, the meta-server was queried to identify targets that are predicted by at least 3 different prediction servers, and the top 30 ranked targets were tabulated (a small fraction of the total number of predictions).
has-miR-20a	E2F1, RUNX1	Smalheiser et al., 2012	Combination of evidence from 11 miRNA target prediction servers using the miRecords meta-server. For each downloaded miRNA, the meta-server was queried to identify targets that are predicted by at least 3 different prediction servers, and the top 30 ranked targets were tabulated (a small fraction of the total number of predictions).
miR-20b, miR-20a, miR-34a, miR-34b	Vascular Endothelial Growth Factor A (VEGFA)	Smalheiser et al., 2012; Dwivedi and Smalheiser, unpublished data in Dwivedi, 2011	Combination of evidence from 11 miRNA target prediction servers using the miRecords meta-server (for more details: http://mirecords.umn.edu/mirecords/; Xiao et al., 2009).
miR-34a	Bcl-2	Dwivedi and Smalheiser, unpublished data in Dwivedi, 2011	Combination of evidence from 11 miRNA target prediction servers using the miRecords meta-server (for more details: http://mirecords.umn.edu/mirecords/; Xiao et al., 2009).
miR-148b	DNMT3b	Dwivedi and Smalheiser, unpublished data in Dwivedi, 2011	Combination of evidence from 11 miRNA target prediction servers using the miRecords meta-server (for more details: http://mirecords.umn.edu/mirecords/; Xiao et al., 2009).
miR-101, miR-34a	MYCN	Dwivedi and Smalheiser, unpublished data in Dwivedi, 2011	Combination of evidence from 11 miRNA target prediction servers using the miRecords meta-server (for more details: http://mirecords.umn.edu/mirecords/; Xiao et al., 2009).

Table 1. (Continued)

Predicted miRNAs	Candidate genes involved in suicide	Author(s), year	Prediction software
miR-195, miR-15a, miR-15b	5-HT2A	Beveridge et al., 2010	Putative target genes were identified using TargetCombo (which combines information shared by multiple databases such as Diana-microT, PicTar, TargetScanS and miRanda available at: http://www.diana. pcbi.upenn.edu/cgi-bin/TargetCombo.cgi). Pathway analysis of the target gene lists were carried out using the Database for Annotation, Visualization and Integrated Discovery bioinformatics resource (http://david.abcc.ncifcrf.gov/).
miR-195	5-HT4	Guo et al., 2010	All the miRNA Target Prediction were derived using the TargetScan server (version 4.2, April 2008). MiRNAs and target gene pairs were extracted by the corresponding miRNA lists (e.g. SZmiRNAs) and genes (e.g. SZgenes).
miR-181b	calcium sensor gene visinin-like 1 (VSNL1); the ionotropic AMPA glutamate receptor subunit (GRIA2)	Beveridge et al., 2008	An extensive list was assembled from a variety of publicly available miRNA databases using the online TargetCombo web service (http://www.diana.pcbi.upenn.edu.cgi-bin/TargetCombo.cgi) with the option to combine target predictions from the leading homology search algorithms miRanda, PicTar,TargetScanS and Diana-microT.

Although few direct evidence primarily investigating the involvement of miRNAs in major depression and suicidal behaviour, a central role of miRNAs in the development of major affective disorders may be hypothesized based on these preliminary studies. The first direct study in literature that primarily investigates the role of miRNAs in patients with major affective disorders and suicidal behaviour is that of Smalheiser et al. (2012). The authors found that miRNA expression was globally down-regulated, by 17% on average, in the prefrontal cortex of 18 depressed suicide victims when compared to that of 17 well-matched non-psychiatric controls died for other causes (specifically, 21 miRNAs were significantly decreased).

An increase of Hsa-miR-185 expression levels have been suggested to regulate the TrkB-T1 decrease in the frontal cortex of 38 subjects who died by suicide who were compared with normal controls (N=17) (Maussion et al, 2012). The authors found that Hsa-miR-185 and Hsa-miR-491-3p were up-regulated in those who completed suicide with a low expression of TrkB-T1. Five possible binding sites for the DiGeorge syndrome linked miRNA Hsa-miR-185 in the 3′UTR of TrkB-T1 emerged at bioinformatic analyses. Also, an inverse relationship between Hsa-miR-185 and trkB-T1 expressions has been confirmed by silencing and overexpression studies in human cell lines.

Indirect contributions also suggest a prominent role of miRNAs in suicidal behaviours. Depressed suicide victims exhibited a down-regulation of protein kinases A (PKA) and PKC (Pandey et al, 1997; Dwivedi and Pandey, 2008) which are responsible of subsequent phosphorylation of transcriptional factors such as CREB. CREB phosphorylation is known to be involved in norepinephrine and 5HT cascades which result associated with depression and presumably suicidal behaviour. Understanding the exact level of this involvement at a molecular level is, no doubt, difficult (Serafini et al, 2011); however, miRNAs are hypothesized to actively participate in these molecular events. Both mRNA and protein levels of nerve growth factor (NGF), neurotrophin 3, and neurotrophin 4/5 (all of similar magnitude) were found significantly decreased in the hippocampus of 28 suicide victims compared with 21 normal controls (Dwivedi et al, 2005a). Table 1 describes the most important studies describing candidate genes involved in suicide and predicted conserved miRNAs that target them via a prediction software.

A reduction of both mRNA and protein levels of neurotrophin 4/5 and NGF in the prefrontal cortex of suicide victims was also found. The decreased amount of these neurotrophins (critically involved in neuronal and glial development, cell survival and functional/structural adult brain integrity) may be explained by a reduced transcription.

Inadequate levels of neurotrophins together with the loss of neurons and glia were associated with the reported structural brain abnormalities (Duman, 2004; Duman et al, 2000).

miRNAs have been also suggested to play a prominent role in suicidal behaviour based on other studies including depressed suicide victims. Dwivedi et al. (2006) reported a reduction of both mRNA level and protein levels of Rap-1 in postmortem brains prefrontal cortex and hippocampus of 28 depressed suicide victims compared with 28 nonpsychiatric control subjects. Reduced transcription may be associated with decreased protein levels of Rap-1 and reduced expression may be related to its reduced activation. No specific abnormalities at a transcriptional level have been hypothesized to play a critical role in the differential regulation of both Rap-1 and Epac (actively involved in neuroprotection and synaptic plasticity). Critical implications for the development and pathophysiology of depression could depend on the substantial alteration of the post-transcriptional regulatory environment characterized by a global increase in miRNA biogenesis and expression in the brain.

Future studies are undoubtedly requested to understand whether post-transcriptional influences on gene expression mediated by changes in ncRNA expression, such as miRNAs, are directly associated with the emergence of suicidal behaviour.

Conclusion

miRNAs presumably play a critical role in synaptic plasticity, neurogenesis, stress responses, major depression and perhaps suicidal behaviour. It has been suggested that various miRNA targets are involved in neural development including mRNAs that encode proteins having a crucial role in the maintenance of neuronal function, plasticity of neural networks, neurodevelopmental disorders (Rogaev, 2005). Interestingly, the expression of specific miRNAs may be restored by the action of some psychoactive drugs such as ADs and mood stabilizers.

In overall, miRNAs may presumably broaden our understanding of the pathophysiology of major affective disorders but several limitations must be considered. First, expression patterns of miRNAs differ in the various brain regions and may be cell-type specific. Also, changes in mature miRNAs expression should be globally analyzed considering that miRNAs regulate either directly and indirectly target mRNAs. In addition, every change detected in mRNAs by gene expression profiling requires independent validation. miRNAs regulation may be performed at multiple levels; eventual deficiencies could result at each of these levels.

The search of miRNA polymorphisms will be determinant in future studies in the effort to discover altered binding to the arrays and altered miRNA expression levels. However, studies aimed to replicate the predicted functional targets of miRNAs associated with depressive phenotypes are also crucial. It is important to consider that some miRNAs or other small non-coding RNAs may not be detected with current technologies. miRNA processing and stability may be affected by specific variations (atypical miRNAs) altering the spectrum of targets that miRNAs regulate and the possible link to neuropsychiatric disorders. Biological processes may be influenced by ncRNAs and differential changes in their expression levels. A relevant role of ncRNAs has been suggested in cognitive functions (Qureshi and Mehler, 2011) whereas other evidence (Vangompel et al., 2009) reported reduced numbers of GABAergic interneurons early in development as well as reduced synaptic inhibition in adulthood in mutant mice lacking Evf2. PIWI-interacting RNA, a small nucleolar ncRNAs, transcribed ultraconserved regions (T-UCRs) and large intergenic ncRNAs are considered as key elements of cellular homeostasis (Esteller, 2011).

Some researchers (Hüttenhofer et al, 2005; Taft et al, 2010) suggested the existence of specific non-coding RNAs such as siRNAs and piRNAs which induced epigenetic regulation through chromatin remodeling events (Verdel et al, 2009). Similarly, additional long piRNAs acting as epigenetic regulators of protein coding gene expression and transcribed by polymerase II promoters or alternatively spliced and/or polyadenylated have been identified. Identifying polymorphisms in post-transcriptional regulatory elements related to miRNAs may give critical information regarding the sequencing of piRNAs and the disease pathophysiology (Sethupathy and Collins, 2008).

Another criticism must be mentioned. It is yet largely unknown how miRNAregulatory network control other cellular signaling networks and how this regulation may contribute in determining biological functions. One of the most critical steps is related to the variability of miRNA targets and the combination of this regulatory network with other signal transduction networks.

It is possible that learning and dendrite spine morphogenesis are dependent by the local control of protein synthesis near synapses and some mature miRNAs are locally processed (being either Dicer and pre-miRNAs expressed in synaptic fractions linked to postsynaptic densities). CREB-BDNF pathways may regulate, through miRNAs, the homeostasis of neural and synaptic pathways playing a prominent role in major depression. However, further studies are required to confirm this hypothesis. This field of research is still rapidly expanding and much remains to be clarified about the role of miRNAs in both major affective disorders and suicidal behaviour.

References

Abelson JF, Kwan KY, O'Roak BJ, Baek DY, Stillman AA, Morgan TM, et al. Sequence variants in SLITRK1 are associated with Tourette's syndrome. *Science,* 2005; 310:317–320.

Alon S, Mor E, Vigneault F, Church GM, Locatelli F, Galeano F, et al. Systematic identification of edited microRNAs in the human brain. *Genome Res.*, 2012;22(8):1533-1540.

Anacker C, Zunszain PA, Cattaneo A, Carvalho LA, Garabedian MJ, Thuret S, et al. Antidepressants increase human hippocampal neurogenesis by activating the glucocorticoid receptor. *Mol. Psychiatry,* 2011; 16 (7): 738–750.

Andersen I, Thielen K, Bech P, Nygaard E, Diderichsen F. Increasing prevalence of depression from 2000 to 2006. *Scand. J., Public Health,* 2011;39:857-863.

Ashraf SI, McLoon AL, Sclarsic SM, Kunes S. Synaptic protein synthesis associated with memory is regulated by the RISC pathway in Drosophila. *Cell,* 2006; 124: 191–205.

Banerjee S, Neveu P, Kosik KS. A coordinated local translational control point at the synapse involving relief from silencing and MOV10 degradation. *Neuron,* 2009; 64: 871–884.

Baudry A, Mouillet-Richard S, Schneider B, Launay JM, Kellermann O. miR-16 targets the serotonin transporter: a new facet for adaptive responses to antidepressants. *Science,* 2010; 329: 1537–1541.

Bekinschtein P, Oomen CA, Saksida LM, Bussey TJ. Effects of environmental enrichment and voluntary exercise on neurogenesis, learning and memory, and pattern separation: BDNF as a critical variable? *Semin. Cell Dev. Biol.,* 2011; 22 (5): 536–542.

Bessa JM, Ferreira D, Melo I, Marques F, Cerqueira JJ, Palha JA, et al. The mood-improving actions of antidepressants do not depend on neurogenesis but are associated with neuronal remodeling. *Mol. Psychiatry,* 2009; 14: 764–773.

Beveridge NJ, Gardiner E, Carroll AP, Tooney PA, Cairns MJ. Schizophrenia is associated with an increase in cortical microRNA biogenesis. *Mol. Psychiatry,* 2010;15:1176–1189.

Beveridge NJ, Tooney PA, Carroll AP, Gardiner E, Bowden N, Scott RJ, et al. Dysregulation of miRNA 181b in the temporal cortex in schizophrenia. *Hum. Mol. Genet.,* 2008; 17: 1156–1168.

Bocchio-Chiavetto L, Maffioletti E, Bettinsoli P, Giovannini C, Bignotti S, Tardito D, et al. Blood microRNA changes in depressed patients during antidepressant treatment. *Eur. Neuropsychopharmacol.,* 2012 (in press).

Brunoni AR, Lopes M, Fregni F. A systematic review and meta-analysis of clinical studies on depression and BDNF levels: implications for the role of neuroplasticity in depression. *Int. J. Neuropsychopharmacol.,* 2008; 11: 1169–1180.

Bushati N, Cohen SM. microRNA functions. *Annu. Rev. Cell Dev. Biol.,* 2007; 23: 175–205.

Castrén E, Rantamäki T. Neurotrophins in depression and antidepressant effects. *Novartis Found. Symp.,* 2008; 289: 43–52.

Castrén E, Voikar V, Rantamäki T. Role of neurotrophic factors in depression. *Curr. Opin. Pharmacol.,* 2007; 7: 18–21.

Caudy AA, Myers M, Hannon GJ, Hammond SM. Fragile X-related protein and VIG associate with the RNA interference machinery. *Genes Dev.,* 2002; 16: 2491–2496.

Chen H, Pandey GN, Dwivedi Y. Hippocampal cell proliferation regulation by repeated stress and antidepressants. *Neuroreport,* 2006; 17: 863–867.

Chen H, Wang N, Burmeister M, McInnis MG. MicroRNA expression changes in lymphoblastoid cell lines in response to lithium treatment. *Int. J. Neuropsychopharmacol.,* 2009; 2: 1–7.

Cheng LC, Pastrana E, Tavazoie M, Doetsch F. miR-124 regulates adult neurogenesis in the subventricular zone stem cell niche. *Nat. Neurosci.,* 2009; 12: 399–408.

Choi PS, Zakhary L, Choi WY, Caron S, Alvarez-Saavedra E, Miska EA, et al. Members of the miRNA-200 family regulate olfactory neurogenesis. *Neuron,* 2008; 57: 41–55.

Christensen M, Schratt GM. microRNA involvement in developmental and functional aspects of the nervous system and in neurological diseases. *Neurosci. Lett.,* 2009; 466: 55–62.

Cimmino A, Calin GA, Fabbri M, Iorio MV, Ferracin M, Shimizu M, et al. *miR-15 and miR-16 induce apoptosis by targeting BCL2. Proc. Natl. Acad. Sci. USA,* 2005; 102: 13944–13949.

Conaco C, Otto S, Han JJ, Mandel G. Reciprocal actions of REST and a microRNA promote neuronal identity. *Proc. Natl. Acad. Sci. USA,* 2006; 103: 2422–2427.

Cooper JD, Skepper JN, Berzaghi MD, Lindholm D, Sofroniew MV. Delayed death of septal cholinergic neurons after excitotoxic ablation of hippocampal neurons during early postnatal development in the rat. *Exp. Neurol.,* 1996; 139: 143–155.

Cowansage KK, LeDoux JE, Monfils MH. Brain-derived neurotrophic factor: a dynamic gatekeeper of neural plasticity. *Curr. Mol. Pharmacol.,* 2010; 3 (1): 12–29.

David DJ, Samuels BA, Rainer Q, Wang JW, Marsteller D, Mendez I, et al. Neurogenesis-dependent and -independent effects of fluoxetine in an animal model of anxiety/depression. *Neuron,* 2009; 62: 479–493.

DeCarolis NA, Eisch AJ. Hippocampal neurogenesis as a target for the treatment of mental illness: a critical evaluation. *Neuropharmacology,* 2010; 58: 884–893.

Duman RS. Introduction: theories of depression—from monoamines to neuroplasticity. In: Olie, JP, Costa e Silva JA, Macher JP eds. *Neuroplasticity: A New Approach to the Pathophysiology of Depression.* London: Science Press Ltd: 1–11, 2004.

Duman RS, Malberg J, Nakagawa S, D'Sa C. Neuronal plasticity and survival in mood disorders. *Biol. Psychiatry,* 2000; 48: 732–739.

Dwivedi Y. Evidence demonstrating role of microRNAs in the etiopathology of major depression. *J. Chem. Neuroanat.,* 2011; 42: 142–156.

Dwivedi Y. Brain-derived neurotrophic factor: role in depression and suicide. *Neuropsychiatr. Dis. Treat.,* 2009;5:433–449.

Dwivedi Y, Rizavi H, Zhang H, Mondal AC, Roberts RC, Conley RR, et al. Neurotrophin receptor activation and expression in human postmortem brain: effect of suicide. *Biol. Psychiatry,* 2009a; 65: 319–328.

Dwivedi Y, Rizavi HS, Zhang H, Roberts RC, Conley RR, Pandey GN. Aberrant extracellular signal-regulated kinase (ERK)1/2 signaling in suicide brain: role of ERK kinase 1 (MEK1). *Int. J. Neuropsychopharmacol.,* 2009b; 12: 1337–1354.

Dwivedi Y, Pandey GN. Adenylyl cyclase-cyclicAMP signaling in mood disorders: role of the crucial phosphorylating enzyme protein kinase A. *Neuropsychiatr. Dis. Treat.,* 2008; 4: 161–176.

Dwivedi Y, Mondal AC, Rizavi HS, Faludi G, Palkovits M, Sarosi A, et al. Differential and brain region-specific regulation of Rap-1 and Epac in depressed suicide victims. *Arch. Gen. Psychiatry,* 2006; 63: 639–648.

Dwivedi Y, Mondal AC, Rizavi HS, Conley RR. Suicide brain is associated with decreased expression of neurotrophins. *Biol. Psychiatry*, 2005a; 58: 315–324.

Dwivedi Y, Mondal AC, Rizavi HS, Shukla PK, Pandey GN. Single and repeated stress-induced modulation of phospholipase C catalytic activity and expression: role in LH behavior. *Neuropsychopharmacology,* 2005b; 30: 473–483.

Dwivedi Y, Mondal AC, Payappagoudar GV, Rizavi HS. Differential regulation of serotonin (5HT)2A receptor mRNA and protein levels after single and repeated stress in rat brain: role in learned helplessness behavior. *Neuropharmacology,* 2005c; 48: 204–214.

Dwivedi Y, Mondal AC, Shukla PK, Rizavi HS, Lyons J. Altered protein kinase a in brain of learned helpless rats: effects of acute and repeated stress. *Biol. Psychiatry,* 2004; 56: 30–40.

Dwivedi Y, Rao JS, Rizavi HS, Kotowski J, Conley RR, Roberts RC, et al. Abnormal expression and functional characteristics of cyclic adenosine monophosphate response element binding protein in postmortem brain of suicide subjects. *Arch. Gen. Psychiatry,* 2003a; 60: 273–282.

Dwivedi Y, Rizavi HS, Conley RR, Roberts RC, Tamminga CA, Pandey GN. Altered gene expression of brain-derived neurotrophic factor and receptor tyrosine kinase B in postmortem brain of suicide subjects. *Arch. Gen. Psychiatry,* 2003b; 60: 804–815.

Esteller M. Non-coding RNAs in human disease. *Nat. Rev. Genet.,* 2011; 12 (12): 861–874.

Fanous S, Hammer Jr, RP, Nikulina EM. Short- and long-term effects of intermittent social defeat stress on brain-derived neurotrophic factor expression in mesocorticolimbic brain regions. *Neuroscience,* 2010; 167: 598–607.

Finkel T, Deng CX, Mostoslavsky R. Recent progress in the biology and physiology of sirtuins. *Nature,* 2009; 460: 587–591.

Fiore R, Siegel G, Schratt G. MicroRNA function in neuronal development, plasticity and disease. *Biochim. Biophys. Acta,* 2008; 1779: 471–478.

Flavell SW, Greenberg ME. Signaling mechanisms linking neuronal activity to gene expression and plasticity of the nervous system. *Annu. Rev. Neurosci.,* 2008; 31: 563–590.

Gao J, Wang WY, Mao YW, Gräff J, Guan JS, Pan L, et al. A novel pathway regulates memory and plasticity via SIRT1 and miR-134. *Nature,* 2010; 466: 1105–1109.

Gao FB. Posttranscriptional control of neuronal development by microRNA networks. *Trends Neurosci.,* 2008; 31: 20–26.

Guo AY, Sun J, Jia P, Zhao Z. A novel microRNA and transcription factor mediated regulatory network in schizophrenia. *BMC Syst. Biol.,* 2010;4:10.

Hansen T, Olsen L, Lindow M, Jakobsen KD, Ullum H, Jonsson E, et al. Brain expressed microRNAs implicated in schizophrenia etiology. PLoS One 2007;2(9):e873.

He L, He X, Lim LP, de Stanchina E, Xuan Z, Liang Y, et al. A microRNA component of the p53 tumour suppressor network. *Nature,* 2007; 447:1130-1134.

Huang YW, Ruiz CR, Eyler EC, Lin K, Meffert MK. Dual regulation of miRNA biogenesis generates target specificity in neurotrophin-induced protein synthesis. *Cell,* 2012;148(5):933-946.

Huang GJ, Herbert J. Stimulation of neurogenesis in the hippocampus of the adult rat by fluoxetine requires rhythmic change in corticosterone. *Biol. Psychiatry,* 2006; 59: 619–624.

Hunsberger JG, Austin DR, Chen G, Manji HK. MicroRNAs in Mental Health: From Biological Underpinnings to Potential Therapies. *Neuromolecular. Med.,* 2009;11:173–182.

Hüttenhofer A, Schattner P, Polacek N. Non-coding RNAs: hope or hype? *Trends Genet.,* 2005; 21: 289–297.

Im HI, Hollander JA, Bali P, Kenny PJ. MeCP2 controls BDNF expression and cocaine intake through homeostatic interactions with microRNA-212. *Nat. Neurosci.,* 2010; 13 (9): 1120–1127.

Impey S, Davare M, Lasiek A, Fortin D, Ando H, Varlamova O, et al. An activity-induced micro- RNA controls dendritic spine formation by regulating Rac1-PAK signaling. *Mol. Cell Neurosci.,* 2010; 43: 146–156.

Ishizuka A, Siomi MC, Siomi H. A Drosophila fragile X protein interacts with components of RNAi and ribosomal proteins. *Genes Dev.,* 2002; 16: 2497–2508.

Jaworski J, Kapitein LC, Gouveia SM, Dortland BR, Wulf PS, Grigoriev I, et al. Dynamic microtubules regulate dendritic spine morphology and synaptic plasticity. *Neuron,* 2009; 61:85–100.

Jensen KP, Covault J, Conner TS, Tennen H, Kranzler HR, Furneaux HM. A common polymorphism in serotonin receptor 1B mRNA moderates regulation by miR-96 and associates with aggressive human behaviors. *Mol. Psychiatry,* 2009; 14: 381–389.

Jin P, Zarnescu DC, Ceman S, Nakamoto M, Mowrey J, Jongens TA, Net al. Biochemical and genetic interaction between the fragile X mental retardation protein and the microRNA pathway. *Nat. Neurosci.,* 2004; 7: 113–117.

Johnson R, Zuccato C, Belyaev ND, Guest DJ, Cattaneo E, Buckley NJ. A microRNA-based gene dysregulation pathway in Huntington's disease. *Neurobiol. Dis.,* 2008;29:438–445.

Kagias K, Nehammer C, Pocock R. Neuronal responses to physiological stress. *Front. Genet.,* 2012;3:222.

Kapsimali M, Kloosterman WP, de Bruijn E, Rosa F, Plasterk RH, Wilson SW. MicroRNAs show a wide diversity of expression profiles in the developing and mature central nervous system. *Genome Biol.*, 2007; 8: R173.

Karege F, Vaudan G, Schwald M, Perroud N, La Harpe R. Neurotrophin levels in postmortem brains of suicide victims and the effects of antemortem diagnosis and psychotropic drugs. *Brain Res. Mol. Brain Res.*, 2005; 136: 29–37.

Kawashima H, Numakawa T, Kumamaru E, Adachi N, Mizuno H, Ninomiya M, et al. Glucocorticoid attenuates brain-derived neurotrophic factor-dependent upregulation of glutamate receptors via the suppression of microRNA-132 expression. *Neuroscience,* 2010; 165: 1301–1311.

Kim AH, Reimers M, Maher B, Williamson V, McMichael O McClay JL. MicroRNA expression profiling in the prefrontal cortex of individuals affected with schizophrenia and bipolar disorders. *Schizophr. Res.*, 2010; 124:183–191.

Kim HW, Rapoport SI, Rao JS. Altered expression of apoptotic factors and synaptic markers in postmortem brain from bipolar disorder patients. *Neurobiol. Dis.*, 2010; 37: 596–603.

Kim J, Inoue K, Ishii J, Vanti WB, Voronov SV, Murchison E, et al. A MicroRNA feedback circuit in midbrain dopamine neurons. *Science,* 2007;317:1220–1224.

Kim J, Krichevsky A, Grad Y, Hayes GD, Kosik KS, Church GM, et al. Identification of many microRNAs that copurify with polyribosomes in mammalian neurons. *Proc. Natl. Acad. Sci. U. S. A.*, 2004;101: 360–365.

Klein ME, Lioy DT, Ma L, Impey S, Mandel G, Goodman RH. Homeostatic regulation of MeCP2 expression by a CREB-induced microRNA. *Nat. Neurosci.*, 2007; 10(12); 1513–1514.

Koehl M, Abrous DN. A new chapter in the field of memory: adult hippocampal neurogenesis. *Eur. J. Neurosci.*, 2011; 33(6):1101–1114.

Konopka W, Kiryk A, Novak M, Herwerth M, Parkitna JR, Wawrzyniak M, et al. MicroRNA loss enhances learning and memory in mice. *J. Neurosci.*, 2010; 30: 14835–14842.

Kosik KS. The neuronal microRNA system. *Nat. Rev. Neurosci.*, 2006; 7: 911–920.

Kozisek ME, Middlemas D, Bylund DB. Brain-derived neurotrophic factor and its receptor tropomyosin-related kinase B in the mechanism of action of antidepressant therapies. *Pharmacol. Ther.*, 2008; 117: 30–51.

Kozomara A, Griffiths-Jones S. miRBase: integrating microRNA annotation and deep-sequencing data. *Nucleic Acids Res,* 2011; 39:D152–D157.

Kumamaru E, Numakawa T, Adachi N, Yagasaki Y, Izumi A, Niyaz M, et al. Glucocorticoid prevents brain-derived neurotrophic factor mediated maturation of synaptic function in developing hippocampal neurons through reduction in the activity of mitogen-activated protein kinase. *Mol. Endocrinol.*, 2008; 22: 546–558.

Kuss AW, Chen W. MicroRNAs in brain function and disease. *Curr. Neurol. Neurosci. Rep.*, 2008; 8(3):190-197.

Landgraf P, Rusu M, Sheridan R, Sewer A, Iovino N, Aravi A, et al. Amammalian microRNA expression atlas based on small RNA library sequencing. *Cell,* 2007; 129: 1401–1414.

Lazarov O, Mattson MP, Peterson DA, Pimplikar SW, van Praag H. When neurogenesis encounters aging and disease. *Trends Neurosci.*, 2010; 33: 569–579.

Lee ST, Chu K, Jung KH, Kim JH, Huh JY, Yoon H, et al. miR-206 regulates brain-derived neurotrophic factor in Alzheimer disease model. *Ann. Neurol.*, 2012;72(2):269-277.

Liu C, Zhao X. MicroRNAs in adult and embryonic neurogenesis. *Neuromol. Med.,* 2009; 11: 141–152.

Lucassen PJ, Meerlo P, Naylor AS, van Dam AM, Dayer AG, Fuchs E, et al. Regulation of adult neurogenesis by stress, sleep disruption, exercise and inflammation: implications for depression and antidepressant action. *Eur. Neuropsychopharmacol.,* 2010; 20: 1–17.

Lugli G, Torvik VI, Larson J, Smalheiser NR. Expression of microRNAs and their precursors in synaptic fractions of adult mouse forebrain. *J. Neurochem.,* 2008; 106: 650–661.

Machado-Vieira R, Salvadore G, DiazGranados N, Ibrahim L, Latov D, Wheeler-Castillo C, et al. New therapeutic targets for mood disorders. *Sci. World J.* 2010; 10: 713–726.

Malan-Müller S, Hemmings SM, Seedat S. Big Effects of Small RNAs: A Review of MicroRNAs in Anxiety. *Mol. Neurobiol.,* 2012 (in press).

Marmol F. Lithium: bipolar disorder and neurodegenerative diseases. Possible cellular mechanisms of the therapeutic effects of lithium. *Prog. Neuropsychopharmacol. Biol. Psychiatry,* 2008; 232: 1761–1771.

Martin KC, Kosik KS. Synaptic tagging—who's it? *Nat. Rev. Neurosci.,* 2002; 3:813–820.

Maussion G, Yang J, Yerko V, Barker P, Mechawar N, Ernst C, et al. Regulation of a truncated form of tropomyosin-related kinase B (TrkB) by Hsa-miR-185* in frontal cortex of suicide completers. *PLoS One,* 2012;7(6):e39301.

McAllister AK, Katz LC, Lo DC. Neurotrophins and synaptic plasticity. *Annu. Rev. Neurosci.,* 1999; 22: 295–318.

McAllister AK. Neurotrophins and neuronal differentiation in the central nervous system. *Cell Mol. Life Sci.,* 2001; 58: 1054–1060.

McEwen BS, Chattarji S. Molecular mechanisms of neuroplasticity and pharmacological implications: the example of tianeptine. *Eur. Neuropsychopharmacol.,* 2004; 14 (Suppl. 5): S497–S502.

McNeill E, Van Vactor D. MicroRNAs Shape the Neuronal Landscape. *Neuron,* 2012; 75: 363-379.

Meerson A, Cacheaux L, Goosens KA, Sapolsky RM, Soreq H, Kaufer D. Changes in brain MicroRNAs contribute to cholinergic stress reactions. *J. Mol. Neurosci.,* 2010; 40 (1–2): 47–55.

Mellios N, Huang HS, Grigorenko A, Rogaev E, Akbarian S. A set of differentially expressed miRNAs, including miR-30a-5p, act as post-transcriptional inhibitors of BDNF in prefrontal cortex. *Hum. Mol. Genet.,* 2011; 17: 3030–3042.

Michalak P. RNA world—the dark matter of evolutionary genomics. *J. Evol. Biol.,* 2006; 19: 1768–1774.

Millan MJ. MicroRNA in the regulation and expression of serotonergic transmission in the brain and other tissues. *Curr. Opin. Pharmacol.,* 2011; 11: 11–22.

Miranda KC, Huynh T, Tay Y, Ang YS, Tam WL, Thomson AM, et al. A pattern-based method for the identification of MicroRNA binding sites and their corresponding heteroduplexes. *Cell,* 2006;126(6):1203-1217.

miRBase. Available from: the web address: http://www.mirbase.org/, 2012.

Miska EA, Alvarez-Saavedra E, Townsend M, Yoshii A, Sestan N, Rakic P, et al. Microarray analysis of microRNA expression in the developing mammalian brain. *Genome Biol.,* 2004; 5: R68.

Mor E, Cabilly Y, Goldshmit Y, Zalts H, Modai S, Edry L, et al. Species-specific microRNA roles elucidated following astrocyte activation. *Nucleic Acids Res.,* 2011;39(9):3710-3723.

Mouillet-Richard S, Baudry A, Launay JM, Kellermann O. MicroRNAs and depression. *Neurobiol. Dis.,* 2012; 46 (2): 272–278.

Nakahata Y, Kaluzova M, Grimaldi B, Sahar S, Hirayama J, Chen D, Guarente LP, et al. The NAD+-dependent deacetylase SIRT1 modulates CLOCK-mediated chromatin remodeling and circadian control. *Cell,* 2008; 134: 329–340.

Nerini-Molteni S, Mennecozzi M, Fabbri M, Sacco MG, Vojnits K, Compagnoni A, et al. MicroRNA Profiling as a Tool for Pathway Analysis in a Human in Vitro Model for Neural Development. *Curr. Med. Chem.,* 2012 (in press).

Newman MA, Hammond SM. Emerging paradigms of regulated microRNA processing. *Genes Dev.,* 2010; 24 (11): 1086–1092.

Numakawa T, Kumamaru E, Adachi N, Yagasaki Y, Izumi A, Kunugi H. Glucocorticoid receptor interaction with TrkB promotes BDNF-triggered PLCgamma signaling for glutamate release via a glutamate transporter. *Proc. Natl. Acad. Sci. U. S. A.,* 2009; 106: 647–652.

O'Connor RM, Dinan TG, Cryan JF. Little things on which happiness depends: microRNAs as novel therapeutic targets for the treatment of anxiety and depression. *Mol. Psychiatry,* 2012; 17 (4): 359–376.

Ohira K, Hayashi M. A new aspect of the TrkB signaling pathway in neural plasticity. *Curr. Neuropharmacol.,* 2009; 7 (4): 276–285.

Otto SJ, McCorkle SR, Hover J, Conaco C, Han JJ, Impey S, Yochum GS, et al. A new binding motif for the transcriptional repressor REST uncovers large gene networks devoted to neuronal functions. *J. Neurosci.,* 2007; 27: 6729–6739.

Ovtscharoff Jr, W, Braun K. Maternal separation and social isolation modulate the postnatal development of synaptic composition in the infralimbic cortex of Octodon degus. *Neuroscience,* 2001; 104: 33–40.

Pandey GN, Dwivedi Y, Pandey SC, Conley RR, Roberts RC, Tamminga CA. Protein kinase C in the postmortem brain of teenage suicide victims. *Neurosci. Lett.,* 1997; 228: 111–114.

Pariante CM, Miller AH. Glucocorticoid receptors in major depression: relevance to pathophysiology and treatment. *Biol. Psychiatry,* 2001; 49: 391–404.

Pascual R, Zamora-León P, Catalán-Ahumada M, Valero-Cabré A. Early social isolation decreases the expression of calbindin D-28k and dendritic branching in the medial prefrontal cortex of the rat. *Int. J. Neurosci.,* 2007; 117: 465–476.

Pathania M, Torres-Reveron J, Yan L, Kimura T, Lin TV, Gordon V, et al. miR-132 enhances dendritic morphogenesis, spine density, synaptic integration, and survival of newborn olfactory bulb neurons. *PLoS One,* 2012;7(5):e38174.

Perkins DO, Jeffries CD, Jarskog LF, Thomson JM, Woods K, Newman MA, et al. microRNA expression in the prefrontal cortex of individuals with schizophrenia and schizoaffective disorder. *Genome Biol.,* 2007;8:R27.

Presutti C, Rosati J, Vincenti S, Nasi S. Non coding RNA and brain. *BMC Neurosci.,* 2006; Suppl 1: S5.

Qurashi A, Chang S, Peng J. Role of microRNA pathway in mental retardation. *Sci. World J.,* 2007; 7: 146–154.

Qureshi IA, Mehler MF. Non-coding RNA networks underlying cognitive disorders across the lifespan. *Trends Mol. Med.,* 2011; 17 (6): 337–346.

Rajasethupathy P, Fiumara F, Sheridan R, Betel D, Puthanveettil SV, Russo JJ, et al. Characterization of small RNAs in Aplysia reveals a role for miR-124 in constraining synaptic plasticity through CREB. *Neuron,* 2009; 63: 803–817.

Rinaldi A, Vincenti S, De Vito F, Bozzoni I, Oliverio A, Presutti C, et al. Stress induces region specific alterations in microRNAs expression in mice. *Behav. Brain Res.,* 2010; 208: 265–269.

Rogaev EI. Small RNAs in human brain development and disorders. *Biochemistry, (Mosc)* 2005; 70: 1404–1407.

Rong H, Liu TB, Yang KJ, Yang HC, Wu DH, Liao CP, et al. MicroRNA-134 plasma levels before and after treatment for bipolar mania. *J. Psychiatr. Res.,* 2011; 45: 92–95.

Sairanen M, O'Leary OF, Knuuttila JE, Castren E. Chronic antidepressant treatment selectively increases expression of plasticity-related proteins in the hippocampus and medial prefrontal cortex of the rat. *Neuroscience,* 2007; 144: 368–374.

Saugstad JA. MicroRNAs as effectors of brain function with roles in ischemia and injury, neuroprotection, and neurodegeneration. *J. Cereb. Blood Flow. Metab.,* 2010; 30: 1564–1576.

Saus E, Brunet A, Armengol L, Alonso P, Crespo JM, Fernández-Aranda F, et al. Comprehensive copy number variant (CNV) analysis of neuronal pathways genes in psychiatric disorders identifies rare variants within patients. *J. Psychiatr. Res.,* 2010; 44: 971–978.

Sempere LF, Freemantle S, Pitha-Rowe I, Moss E, Dmitrovsky E, Ambros V. Expression profiling of mammalian microRNAs uncovers a subset of brain-expressed microRNAs with possible roles in murine and human neuronal differentiation. *Genome Biol.,* 2004; 5 (3): R13.

Serafini G, Pompili M, Innamorati M, Giordano G, Montebovi F, Sher L, et al. The role of microRNAs in synaptic plasticity, major affective disorders and suicidal behavior. *Neurosci. Res.,* 2012;73(3):179-190.

Serafini G, Pompili M, Innamorati M, Giordano G, Tatarelli R, Lester D, et al. Glycosides, depression and suicidal behaviour: the role of glycoside-linked proteins. *Molecules,* 2011; 16(3):2688-2713.

Sethupathy P, Collins FS. MicroRNA target site polymorphisms and human disease. *Trends Genet.,* 2008; 24: 489–497.

Schaeffer C, Beaulande M, Ehresmann C, Ehresmann B, Moine H. The RNA binding protein FMRP: new connections and missing links. *Biol. Cell.,* 2003; 95:221-228.

Schratt GM, Tuebing F, Nigh EA, Kane CG, Sabatini ME, Kiebler M, et al. A brain-specific microRNA regulates dendritic spine development. *Nature,* 2006; 439: 283–289.

Shi SS, Shao SH, Yuan BP, Pan F, Li ZL. Acute stress and chronic stress change brain-derived neurotrophic factor (BDNF) and tyrosine kinase-coupled receptor (TrkB) expression in both young and aged rat hippocampus. *Yonsei Med. J.,* 2010; 51: 661–671.

Siegel G, Obernosterer G, Fiore R, Oehmen M, Bicker S, Christensen M, et al. A functional screen implicates microRNA-138-dependent regulation of the depalmitoylation enzyme APT1 in dendritic spine morphogenesis. *Nat. Cell. Biol.,* 2009; 11: 705–716.

Smalheiser NR, Lugli G, Rizavi HS, Torvik VI, Turecki G, Dwivedi Y. MicroRNA expression is down-regulated and reorganized in prefrontal cortex of depressed suicide subjects. *PLoS One,* 2012;7(3):e33201.

Smirnova L, Gräfe A, Seiler A, Schumacher S, Nitsch R, Wulczyn FG. Regulation of miRNA expression during neural cell specification. *Eur. J. Neurosci.,* 2005; 21 (6): 1469–1477.

Smith MA, Makino S, Kvetnansky R, Post RM. Stress and glucocorticoids affect the expression of brain-derived neurotrophic factor and neurotrophin-3 mRNAs in the hippocampus. *J. Neurosci.,* 1995; 15: 1768–1777.

Sofroniew MV, Galletly NP, Isacson O, Svendsen CN. Survival of adult basal cholinergic neurons after loss of target neurons. *Science,* 1990; 247: 338–342.

Soldati C, Bithell A, Johnston C, Wong KY, Stanton LW, Buckley NJ. Dysregulation of REST-regulated coding and non-coding RNAs in a cellular model of Huntington's disease. *J. Neurochem.,* 2012 (in press).

Stevenson CW, Marsden CA, Mason R. Early life stress causes FG-7142- induced corticolimbic dysfunction in adulthood. *Brain Res.,* 2008; 1193: 43–50.

Szulwach KE, Li X, Smrt RD, Li Y, Luo Y, Lin L, et al. Cross talk between microRNA and epigenetic regulation in adult neurogenesis. *J. Cell Biol.,* 2010; 189: 127–141.

Taft RJ, Pang KC, Mercer TR, Dinger M, Mattick JS. Non-coding RNAs: regulators of disease. *J. Pathol.,* 2010; 220: 126–139.

Thoenen H. Neurotrophins and activity-dependent plasticity. *Prog. Brain Res.,* 2000; 128: 183–191.

Turner JD, Alt SR, Cao L, Vernocchi S, Trifonova S, Battello N, et al. Transcriptional control of the glucocorticoid receptor: CpG islands, epigenetics and more. *Biochem. Pharmacol.,* 2010; 80: 1860–1888.

Uchida S, Nishida A, Hara K, Kamemoto T, Suetsugi M, Fujimoto M, et al. Characterization of the vulnerability to repeated stress in Fischer 344 rats: possible involvement of microRNA-mediated down-regulation of the glucocorticoid receptor. *Eur. J. Neurosci.,* 2008; 27: 2250–2261.

Uchida S, Hara K, Kobayashi A, Funato H, Hobara T, Otsuki K, et al. Early life stress enhances behavioral vulnerability to stress through the activation of REST4-mediated gene transcription in the medial prefrontal cortex of rodents. *J. Neurosci.,* 2010; 30, 15007–15018.

Vangompel MJ, Sametsky EA, Clark MF, Savage JC, Disterhoft JF, Kohtz JD, et al. Balanced gene regulation by an embryonic brain ncRNA is critical for adult hippocampal GABA circuitry. *Nat. Neurosci.,* 2009; 12: 1020–1027.

Verdel A, Vavasseur A, Le Gorrec M, Touat-Todeschini L. Common themes in siRNA-mediated epigenetic silencing pathways. *Int. J. Dev. Biol.,* 2009; 53; 245–257.

Vo N, Klein ME, Varlamova O, Keller DM, Yamamoto T, Goodman RH, et al. A cAMP-response element binding protein-induced microRNA regulates neuronal morphogenesis. *Proc. Natl. Acad. Sci. U. S. A.,* 2005; 102, 16426–16431.

Vreugdenhil E, Berezikov E. Fine-tuning the brain: microRNAs. *Front. Neuroendocrinol.,* 2009; 31: 128–133.

Vreugdenhil E, Verissimo CS, Mariman R, Kamphorst JT, Barbosa JS, Zweers T, et al. MicroRNA 18 and 124a down-regulate the glucocorticoid receptor: implications for glucocorticoid responsiveness in the brain. *Endocrinology,* 2009; 150, 2220–2228.

Wayman GA, Davare M, Ando H, Fortin D, Varlamova O, Cheng HY, et al. An activity regulated microRNA controls dendritic plasticity by down-regulating p250GAP. *Proc. Natl. Acad. Sci. U. S. A.,* 2008; 105: 9093–9098.

Witke W. The role of profilin complexes in cell motility and other cellular processes. *Trends Cell. Biol.,* 2004; 14: 461–469.

World Health Organization. *Information on mental disorders management: depression.* Geneva: World Health Organization. Available from: http://www.who.int/ mental_ health/management/ depression/definition/ en/index.html, 2012a.

World Health Organization. *Mental Health: suicide prevention.* Geneva: World Health Organization. Available from: http://www.who.int/mental_ health/ prevention/en/, 2012b.

Wu J, Xie X. Comparative sequence analysis reveals an intricate network among REST, CREB and miRNA in mediating neuronal gene expression. *Genome Biol.,* 2006; 7: R85.

Xiao F, Zuo Z, Cai G, Kang S, Gao X, Li T. miRecords: an integrated resource for microRNA-target interactions. *Nucleic Acids Res.,* 2009;37(Database issue):D105-10.

Xu Y, Liu H, Li F, Sun N, Ren Y, Liu Z, et al. A polymorphism in the microRNA-30e precursor associated with major depressive disorder risk and P300 waveform. *J. Affect. Disord.,* 2010; 127: 332–336.

Yagasaki Y, Numakawa T, Kumamaru E, Hayashi T, Su TP, Kunugi H. Chronic antidepressants potentiate via sigma-1 receptors the brain-derived neurotrophic factor-induced signaling for glutamate release. *J. Biol. Chem.,* 2006;281:12941–12949.

Yirmiya R, Goshen I. Immune modulation of learning, memory, neural plasticity and neurogenesis. *Brain Behav. Immun.,* 2011; 25 (2): 181–213.

Zhou R, Yuan P, Wang Y, Hunsberger JG, Elkahloun A, Wei Y, et al. Evidence for selective microRNAs and their effectors as common long-term targets for the actions of mood stabilizers. *Neuropsycho-pharmacology,* 2009; 34: 1395–1405.

Zhou Y, Ferguson J, Chang JT, Kluger Y. Inter- and intra-combinatorial regulation by transcription factors and microRNAs. *BMC Genomics,* 2007; 8: 396.

In: MicroRNA and Non-Coding RNA
Editor: James C. Johnson
ISBN: 978-1-62618-442-8

Chapter 3

MicroRNAs in Endoparasites

Mara Rosenzvit, Marcela Cucher, Laura Kamenetzky, Natalia Macchiaroli, Laura Prada and Federico Camicia
Instituto de Microbiología y Parasitología Médica,
Universidad de Buenos Aires-Consejo Nacional
de Investigaciones Científicas y Tecnológicas (IMPaM, UBA-CONICET),
Facultad de Medicina, Buenos Aires, Argentina

Abstract

MicroRNAs (miRNAs) are short, non-coding RNAs that down-regulate gene expression post-transcriptionally by binding to the mRNA of their target genes. The importance of miRNAs in key biological processes such as development, cell proliferation, cell differentiation and metabolism has been widely documented since their discovery. To date, more than 25,000 mature miRNAs have been reported in 193 species of animals and plants (miRBase, release 19, http://www.mirbase.org /index.shtml). Parasites face different environments during their complex life cycles and their ability to respond to environmental and developmental signals may require miRNA-mediated networking and fine-tuning of gene expression. Some protozoan and helminth parasites have shown peculiarities in their small-RNA machinery. For example, the small RNA-generating machinery of the apicomplexan parasite *Toxoplasma gondii* is phylogenetically and functionally related to that of plants and fungi. Among protozoan parasites, miRNAs have been reported in *Giardia lamblia*, *Trichomonas vaginalis*, *Trichomonas foetus*, *Pentatrichomonas hominis*, *Trypanosoma brucei*, *Entameoba histolytica*, *Neospora caninum* and *Toxoplasma gondii*. Most of the protozoan miRNAs do not share significant homology to any of the known miRNAs of plants and metazoans. Some interesting roles such as silencing of variant surface proteins have been proposed for a number of miRNAs in *G. lamblia*. However, in other unicellular parasites such as *Trypanosoma cruzi* and *Plasmodium falciparum* no miRNAs or small RNA machinery components have been identified. Among helminth parasites, miRNA identification and expression profile analyses have been performed by a combination of bioinformatic and experimental approaches. So far, miRNAs have been reported in the nematodes *Angiostrongylus cantonensis*, *Brugia pahangi*, *Brugia malayi*, *Haemonchus contortus*, *Ascaris suum* and *Trichinella spiralis;* the trematodes *Fasciola hepatica*, *Fasciola gigantica, Schistosoma mansoni, Schistosoma japonicum, Clonorchis sinensis and Orientobilharzia turkestanicum* and the cestodes *Echinococcus granulosus, Echinococcus multilocularis* and *Taenia saginata*. Conserved as well as novel miRNAs have been reported, some of them are abundantly expressed and display developmental regulation, suggesting important roles in several helminth parasite species. miRNA targets have been bioinformatically predicted in parasitic protozoa and helminths but only a small number of protozoan miRNAs has been experimentally validated. Unraveling miRNA roles in parasite biology and host interaction remains a challenge. For this end, bioinformatic strategies adapted to each phylum together with experimental strategies such as miRNA silencing and/or overexpression should be set up. The knowledge of miRNA roles in parasite development and survival in their hosts will in turn provide new control and intervention strategies against the worldwide distributed and mostly neglected diseases they produce.

1. Introduction

Protozoan and helminth parasites are a major cause of diseases affecting public health and economy worldwide. Protozoan parasites include *Giardia lamblia*, the deeply branching unicellular, flagellated protozoa responsible for water-borne outbreaks of diarrhoea (Morrison et al., 2007); the apicomplexans *Plasmodium falciparum* and *Toxoplasma gondii*, causative agents of malaria and toxoplasmosis respectively (Braun et al., 2010); the trypanosomatids *T. cruzi* and *T. brucei* (Mallick et al., 2008), ethiological agents of Chagas disease and African sleeping sickness respectively, among others. "Helminth parasites" is an operational term that refers to parasitic worms with metazoan origin that infect humans (Jenkins & Allen, 2010). They comprise parasites belonging to Nematoda and Platyhelminthes which split from each other over 1 billion years ago, long predating the split between vertebrates and invertebrates (Jenkins & Allen, 2010 and references therein). The phylum Nematoda (roundworms) include hookworms, whipworms, and filarial parasites and the phylum Platyhelminthes (flatworms) include the cestodes (tapeworms) and trematodes (flukes). Parasitic nematodes represent an enormous burden in terms of human, animal, and even plant health. In terms of human disease, platyhelminths infect fewer numbers but are responsible for higher levels of morbidity and mortality.

miRNAs are a class of small ~22 nucleotides (nt) non-coding RNAs with a major role in regulation of gene expression. miRNAs down-regulate gene expression post-transcriptionally by binding to the mRNA of their target genes and promoting their cleavage, or more commonly in metazoans, their translational repression and/or destabilization. The importance of miRNAs in key biological processes such as development, cell proliferation, cell differentiation and metabolism has been widely documented since their discovery (Bartel, 2004). To date, more than 25,000 mature miRNAs have been reported in 193 species of animals and plants. Each miRNA is able to bind and regulate hundreds of target mRNAs. It has been shown that some miRNAs such as miR-1 or miR-124 are able to change the whole cellular mRNA profile thus defining cell fate (Lim et al., 2005). These features suggest that miRNAs are master regulators of gene expression. The regulation exerted by a miRNA on each gene is often subtle. Also, each mRNA can be regulated by more than one miRNA. As a consequence of all these features miRNAs are considered to perform fine-tune regulation of gene expression allowing a rapid response to environmental changes.

Fine-tune regulation of gene expression is necessary for protozoan and helminth parasites in order to count with proper protein repertoires to invade their hosts, be able to survive to the host immune response, migrate and establish in the host, and to develop into very different life cycle stages. Thus, miRNA regulation of gene expression could be very important for parasite development and adaptation to their host/s.

Here we will discuss how endoparasite miRNAs are generated and which proteins are involved in their pathways. We will also review miRNA discovery in parasites, and finally discuss the first attempts to elucidate the role of miRNAs in parasitism

2. miRNA Machinery in Endoparasites

2.1. miRNA Biogenesis in Metazoa

Canonical animal miRNAs are generated from long hairpin precursors (pri-miRNAs) transcribed by RNApol II, through consecutive cleavages by two RNase III enzymes.

In the nucleus, the single strand-double strand junction of the pri-miRNA hairpin is recognized by DGCR8 (also known as Pasha), which positions the catalytic site of the RNase III enzyme Drosha. This cleavage generates a ~70 nt long pre-miRNA hairpin that is exported to the cytoplasm (by Exportin-5), where it is cleaved towards the terminal loop end by the RNase III enzyme Dicer, generating a duplex comprising the mature miRNA and its partially complementary strand (miRNA*).

The mature miRNA is assembled into an effector miRNA Induced Silencing Complex (miRISC) which contains one member of the Argonaute (Ago) protein family.

The miRISC loaded miRNA then associates with a target sequence which is generally in the 3' untranslated region (UTR) of target mRNAs. Depending on the degree of complementarity between the miRNA and the target sequence, either endonucleolytic cleavage (perfect complementarity) or destabilization or translation inhibition of the target mRNA (partial complementarity) can occur.

In animals, the later mechanism is the most frequent and most miRNAs bind their targets only through nt 2–8, which are called the seed sequence (Lewis et al., 2003), although some exceptions were shown (see Thomas et al., 2010).

Two additional pathways for miRNA biogenesis have been described so far in Metazoa apart from the above mentioned canonical pathway (Drosha/DGCR8-Dicer dependent): the Drosha/DGCR8 independent-Dicer dependent and the Drosha/DGCR8 dependent-Dicer independent pathways (Miyoshi et al., 2010, Yang et al., 2011), see Figure 1. In the Drosha/DGCR8 independent-Dicer dependent pathways pre-miRNA-like hairpins that serve as Dicer substrates can be generated by at least four mechanisms:

I. **Mirtron pathway:** mirtrons are short intronic hairpins located in introns of protein coding genes (host genes). They are excised by splicing and linearized by a lariat debranching enzyme, and then the hairpin structure is adopted (Miyoshi et al., 2010).

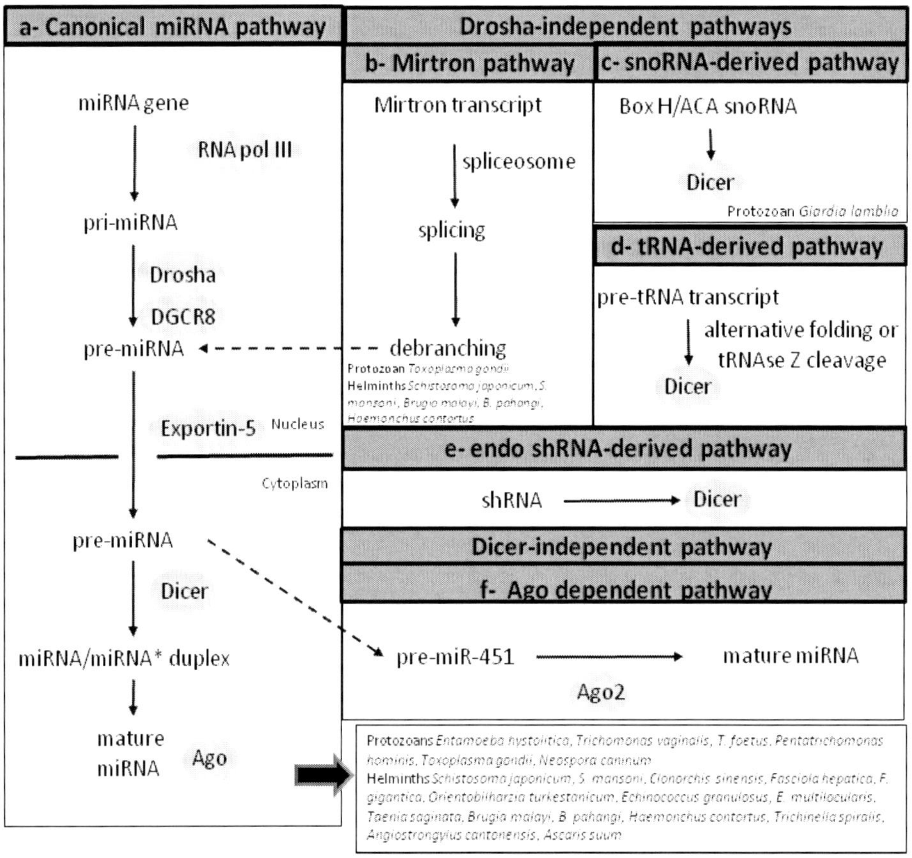

Figure 1. Canonical and non-canonical pathways of miRNA biogenesis based on Miyoshi et al. (2010). a. Canonical pathway of miRNA biogenesis in animals. miRNAs are transcribed from the genome as long primary transcripts. Pri-miRNAs form hairpin structures, and are processed to about 60–70 nt precursor miRNAs (pre-miRNAs) by the Microprocessor (Drosha-DGCR8 complex) in the nucleus. After being exported by Exportin-5, pre-miRNAs are further processed to aproximatly 22 nt miRNA/miRNA*duplex by the Dicer in the cytoplasm. Finally, mature miRNAs are loaded onto Argonaute (Ago) proteins. b. Mirtron derived pathway. Short introns with hairpin potential, termed mirtrons, can be spliced and debranched into pre-miRNA hairpin mimics that bypass Drosha cleavage. Debranched mirtrons access the canonical miRNA pathway during nuclear export, and are then cleaved by Dicer and loaded onto Ago proteins. c. snoRNA-derived pathway. The processing of snoRNAs to miRNAs requires Dicer activity but is independent of Drosha/DGCR8. d. tRNA-derived pathway. pre-tRNA transcript with typical cloverleaf structure can adopt a secondary hairpin like structure or one arm of the pre-tRNA transcript with pre-miRNA-like stem loop structure is processed by tRNase Z. After nuclear export, pre-miRNA is processed by Dicer. e. shRNA-derived pathway. The processing may be similar to that of the snoRNA-derived pathway. f. Dicer-independent pathway. Although Drosha cleavage proceeds normally, the Dicer step is skipped and the pre-mir-451 is loaded directly into Ago2. Parasite organisms with evidence of each pathway are indicated at the bottom of the panels.

II. tRNA-derived pathway: some RNA polymerase III-transcribed tRNAs, which typically form cloverleaf secondary structures, have the potential to form a hairpin as an alternative secondary structure (Babiarz et al., 2008). Also, it has been described that one arm with pre-miRNA-like stem loop structure, located 3' in the tRNA could be processed into pre-miRNA-like hairpins by RNaseZ cleavage (Bogerd et al., 2010). Both tRNA-derived hairpins are substrates for Dicer.

III. Endo-short hairpin pathway: RNAs lacking lower stems present in pri-miRNAs may derive from Pol III transcription or cleavage by as yet unknown endo- or exonucleases (Babiarz et al., 2008).

IV. snoRNA-derived pathway: a number of box C/D and box H/ACA snoRNAs can also give rise to Ago-associated miRNA-like species in a Drosha/DGCR8-independent, Dicer-dependent manner.

The Drosha/DGCR8 dependent-Dicer independent pathway was described for miR-451, a functional miRNA that is perfectly conserved among vertebrates (Yang et al., 2010). Pre-mir-451 produced by Drosha is directly loaded onto Ago2 without the Dicer step, and the Ago2 catalytic centre helps to catalyze the maturation of this miRNA.

Dicer and Ago proteins are central to miRNA biogenesis and function in all species analyzed so far, but certain homologs have distinct properties (Ghildiyal & Zamore, 2009). For example, *Caenorhabditis elegans* and vertebrates encode a single

Dicer that generates both miRNAs and small interfering RNAs (siRNAs), but *Drosophila* encodes two Dicers, of which Dcr1 is specialized for pre-miRNA cleavage and Dcr2 is selective for siRNA biogenesis. Argonaute proteins also exhibit specialization. *Drosophila* has two Ago-class effectors, of which Ago1 is dominant for miRNAs and Ago2 for siRNAs. However, vertebrates have four Ago-class effectors that participate in miRNA-mediated regulation and only Ago2 has slicer activity (Suh & Blelloch, 2011).

2.2. Proteins Involved in Generation and Function of miRNAs in Endoparasites

Interestingly, miRNA pathways in protozoan and helminth parasites show common as well as unique features like number and class of proteins for miRNA biogenesis and function.

2.2.1. miRNA Machinery in Protozoan Parasites

The miRNA machinery is well known in the protozoans *G. lamblia* and *T. gondii.* Eventhough no homologue of Drosha/Pasha or Exportin 5 has been found in *Giardia* (Saraiya & Wang, 2008) an Argonaute homologue (GlAgo) and a Dicer homologue (GlDcr) were identified and found to be functional in this organism (Macrae et al., 2006; Prucca et al., 2008; Saraiya & Wang, 2008; Li et al., 2011; Saraiya et al., 2011).

The three dimensional structure of GlDcr was resolved by X-ray crystallography and it was shown that it can generate small RNA products in the range of 25-27 nt (Macrae et al., 2006). GlAgo was found to bind miRNAs and be necessary for miRNA-dependent repression of gene expression. Since slicing activity was not found for GIAgo, it was suggested that it could act through repression of translation (Saraiya et al., 2011). The *Giardia* miRNA repertoire includes miRNAs derived from snoRNAs, by a Drosha independent pathway, accordingly with the enzymes found (Saraiya & Wang 2008, Li et al., 2012).

In humans, many miRNA precursors and mature miRNAs have also been detected in the nucleolus (Scott et al., 2009, Ono et al., 2011) and snoRNA-derived miRNAs have been since then demonstrated in a variety of organisms (Taft et al., 2009).

Table 1. microRNA data available for protozoan parasites

Parasite species	Phylum	Identification strategy	miRBase deposition[a]	Available sequences[b]	Northern Blot validation	Reference genome used	Reference[c]
Entamoeba histolytica	Sarcomastigophora	BP	NO	YES	NO	*E. histolytica*	De et al., 2006
Trypanosoma brucei	Sarcomastigophora	BP	NO	NO	NO	*T. brucei*	Mallick et al., 2008
Giardia lamblia	Sarcomastigophora	BP, HS	NO	Partial	YES	*Giardia*	Saraiya and Wang, 2008; ; Chen et al., 2009; Zhang et al., 2009; Saraiya and Wang, 2011; Li et al., 2011; Huang et al., 2012; Li et al., 2012
Trichomonas vaginalis	Sarcomastigophora	BP, HS	NO	Partial	NO	*Trichomonas*	Chen et al., 2009; Lin et al., 2009a; Lin et al., 2009b; Huang et al., 2012
Trichomonas foetus	Sarcomastigophora	HS	NO	NO	NO	N/A	Huang et al., 2012
Pentatrichomonas hominis	Sarcomastigophora	HS	NO	NO	NO	N/A	Huang et al., 2012
Toxoplasma gondii	Apicomplexa	HS	NO	YES	YES	*T. gondii*	Braun et al., 2010; Wang et al., 2012a
Neospora caninum	Apicomplexa	BC	NO	YES	YES	*N. caninum*	Braun et al., 2010

[a]: Release 19.

[b]: Partial: the complete set of sequences described in one or more of the articles has not been reported.

[c]: When more than one reference is available, the one/s in bold meets the criteria of both reported sequences and northern blot validation.

BP: bioinformatic prediction de novo, BC: bioinformatic comparison with related organism genome, HS: high scale sequencing. N/A: not applicable. The sequences were mapped to RNA family database and miRBase.

Table 2. microRNA data available for helminth parasites

Parasite species	Phylum	Identification strategy	miRBase deposition[a]	Available sequences	Northern Blot validation	Reference genome used	Reference[c]
Schistosoma japonicum	Platyhelminthes (Trematoda)	LS, HS, BC	YES	YES	YES	*S. japonicum*	Xue et al., 2008b; Huang et al., 2009; Copeland et al., 2009; Hao et al., 2010; Wang et al., 2010; Cai et al., 2011
Schistosoma mansoni	Platyhelminthes (Trematoda)	BP, BC, LS	YES	YES	YES	*S.japonicum, S. mansoni*	Xue et al., 2008b; Huang et al., 2009; Copeland et al., 2009; de Souza Gomes et al., 2011; Simoes et al., 2011
Clonorchis sinensis	Platyhelminthes (Trematoda)	HS	NO	YES	NO	*S. japonicum*	Xu et al., 2010
Fasciola hepatica	Platyhelminthes (Trematoda)	HS	NO	YES	NO	*S. japonicum*	Xu et al., 2012
Fasciola gigantica	Platyhelminthes (Trematoda)	HS	NO	YES	NO	*S. japonicum*	Xu et al., 2012
Orientobilharzia turkestanicum	Platyhelminthes (Trematoda)	HS	NO	Partial	NO	*S. japonicum*	Wang et al., 2012b
Echinococcus granulosus	Platyhelminthes (Cestoda)	LS	YES	YES	YES	*E. granulosus, E. multilocularis*	Cucher et al., 2011
Echinococcus multilocularis	Platyhelminthes (Cestoda)	BC	YES	YES	NO	*E. multilocularis*	Cucher et al., 2011
Taenia saginata	Platyhelminthes (Cestoda)	HS	NO	NO	NO	*S. japonicum*	Ai et al., 2012
Brugia malayi	Nematoda	BP, LS	YES	YES	YES	*B. malayi*	Poole et al., 2010
Brugia pahangi	Nematoda	HS, BC	NO	YES	NO	*B. malayi*	Winter et al., 2012
Haemonchus contortus	Nematoda	HS, BC	YES	YES	NO	*H. contortus*	Winter et al., 2012
Trichinella spiralis	Nematoda	HS	NO	Partial	YES	*T. spiralis*	Liu et al., 2011, Chen et al., 2011a
Angiostrongylus cantonensis	Nematoda	HS	NO	NO	NO	*C. elegans*	Chen et al., 2011b
Ascaris suum	Nematoda	HS	YES	Partial	YES	*A. suum*	Xu et al., 2013, Wang et al., 2011

[a]: Release 19. Not all the articles may have sequences deposited at miRBase when more than one article is cited for a given organism.
[b]: Partial: the complete set of sequences described in the article or one of the articles has not been reported.
[c]: When more than one reference is available, the one/s in bold meets the criteria of both reported sequences and Northern blot validation.
BP: bioinformatic prediction de novo, BC: bioinformatic comparison with related organism genome, HS: high scale sequencing, LS: low scale cloning.

It could be that during parasitic evolution, *Giardia* prioritized the Drosha/Pasha independent miRNA biogenesis pathway, perhaps due to fewer enzymes and regulation requirements in comparison to the canonical pathway.

Sequence analyses showed that the *T. gondii* genome encodes one Dicer (TgDicer) and one Ago (TgAgo) proteins (Gajria et al., 2008), which display significant variability in primary sequence and domain organization compared to consensus (Braun et al., 2010).

TgDicer possesses an RNA helicase domain and two RNaseIII catalytic domains, but it lacks both recognizable domains for dsRNA binding, DSRM and PAZ domains.

This organization is strikingly reminiscent of the DCL1 protein of the single cell algae *Chlamydomonas reinhardtii*. Also, phylogenetic analysis showed that *Toxoplasma* and *Chlamydomonas* Dicer-like sequences seem to be orthologous. The expression of the TgAgo is supported by a number of tachyzoite cDNA sequences and exhibits unique features.

TgAgo is smaller than reported Ago proteins derived from higher eukaryotic organisms, but has a similar size to those from archaeal bacteria. TgAgo contains a conserved PIWI domain and non-conserved PAZ domain and lack the amino acids required for slicer activity. Also, it is mainly localized in the cytoplasm (Al Riyahi et al., 2006). Furthermore, phylogenetic analysis showed that TgAgo belongs to the Ago-like family but only with weak bootstrap support, suggesting that the protein diverges significantly from its metazoan and plant counterparts (Braun et al., 2010).

For other deep-branching unicellular parasites, such as *Entamoeba histolytica* and *Trichomonas vaginalis*, miRNA sequences were reported (see Table 1) as well as Dicer and Ago homologues which were found by a comparative genomics approach (De et al., 2006; Krautz-Peterson & Skelly 2008). However, no genes coding for Ago and Dicer-like proteins have been found in the genomes of *Leishmania major*, *Leishmania infantum*, *T. cruzi*, *Plasmodium* spp., *Cryptosporidium* spp., *Theileria* spp., *Babesia bovis*, and *Eimeria tenella* (Militello et al., 2008). Recently, the absence of miRNAs in *Plasmodium falciparum* has been experimentally validated by small RNA sequencing, which is in coincidence with the absence of Ago/Dicer genes (Xue et al., 2008a; Rathjen et al., 2006). Thus, not all protozoan parasites have the miRNA regulating pathway.

2.2.2. miRNA Machinery in Helminth Parasites

Schistosoma spp. parasites are among the most studied helminth parasites. The presence of three protein-coding genes encoding crucial components of the miRNA processing machinery (Dicer, Drosha, and Pasha/DGCR8) (Artzi et al., 2008) and the four Ago-like genes in both *S. japonicum* (Liu, 2006) and *S. mansoni* (Copeland et al., 2009) detected by tBLASTn in EST data, strongly suggest that schistosomes have a functional miRNA system. Interestingly, expression patterns studies of the Dicer-like gene during different life cycle stages of *S. mansoni* showed a threefold higher expression level in seven days old schistosomula in comparison to the adult worms (Krautz-Peterson & Skelly, 2008). In the parasitic nematode *Ascaris*, cDNA sequences encoding 10 Argonaute proteins of the Ago and Worm Argonaute (Wago) clades as well as Dicer, Drosha and Pasha were identified (Wang et al., 2011). Also, the *B. malayi* genome contains many of components required for miRNA processing including orthologues of Drosha, Dicer and Argonaute (Scott et al., 2012).

Recently, comparative genomics of four flatworm species, *Echinococcus granulosus*, *Echinococcus multilocularis*, *Hymenolepis microstoma* and *Taenia solium* revealed that, besides the canonical Agos, they have a new clade significantly divergent to other already described Argonaute clades (Tsai et al., in press; Zheng, 2013). These proteins are absent from free living flatworms such as *Schmidtea mediterranea*, suggesting they could have a role in parasitism, perhaps binding yet unknown small RNAs.

The parasite miRNA pathways are shown in Figure 1.

3. miRNA Discovery in Endoparasites

3.1. Tools for miRNA Identification

miRNAs are identified either by bioinformatic approaches, followed by experimental validation of the predicted candidate miRNAs, or by low scale cloning or high throughput sequencing strategies both of which also need bioinformatic tools for the analyses of the obtained sequences. Some of the strategies developed to identify miRNAs are briefly introduced below.

I. **RNA conformation based approaches:** Mfold (Zuker, 2003) and RNAfold (Hofacker, 2003) are folding algorithms for the prediction of the most likely secondary structure of single stranded RNA sequences of putative pre-miRNA. Computation algorithms relying on these programs, such as miRseeker and MiRscan, have been developed to identify animal miRNA genes (Sun et al., 2010).

II. **Homology based methods:** using known miRNA and pre-miRNA sequences, these methods can identify novel miRNAs whose sequences are evolutionarily conserved among different species. ESTs and genome *databases* can be screened using BLASTn tools to identify miRNA and pre-miRNA homologs from miRBase. One potential drawback of these homology based methods is their inability to identify totally novel miRNA sequences due to the conservation criteria on which they rely.

III. **Machine-learning techniques:** several new methods for miRNA prediction are based on properties and features of known miRNAs. Among them, machine-learning techniques use sequence composition and structural conformation features to train a learning system algorithm such that it is capable to identify candidate miRNAs (Saetrom & Snove, 2007). As opposed to homology based methods, this approach could be useful for species-specific miRNA discovery since it does not depend on evolutionary conservation.

IV. **Small scale cloning and high-scale sequencing:** low scale cloning procedures for miRNA identification were first described by Lagos-Quintana et al. (2001) and Lau et al. (2001). The procedure includes the isolation of small RNAs (~18-30 nt), ligation of specialized linkers to the 5' and 3' ends, reverse transcription and PCR amplification. The PCR products obtained are cloned and sequenced. Data analysis consists in mapping sequences to the genome of the same species or the closest one available. Then, sequences not mapping the genome, or homologous to ESTs, rRNAs, tRNAs, snoRNA and other small non-coding RNAs are discarded as they are regarded as turnover products. However, there are exceptions since for some protozoan parasites snoRNA and ORF-derived miRNAs were reported (see section

3.2). The remaining sequences are considered putative miRNAs if they fulfill the following requirements: (i) length between 19–30 nt and (ii) predicted stem–loop structure with a free folding energy lower than -20 kcal/mol with the putative miRNA located on the stem, using the above mentioned folding programs. The putative mature miRNAs are compared with the miRNA database miRBase (http://www.mirbase.org/search.shtml) to analyze their similarity to already reported sequences. In the case that genome data from other species was used, it is suggested to clone and analyze the sequence sorrounding the putative miRNA from the original genome to search for pre-miRNA secuences or to validate the candidate miRNAs by Northern blot.

Next-generation sequencing platform *(also referred* to as deep sequencing), which has improved technology to sequence shorter fragments of DNA/RNA (Roche 454, Illumina, SoLiD) is also used to discover miRNAs. The sequence data analysis is similar to the above mentioned strategy and some bioinformatics tools are already available (http://centre.bioinformatics.zj.cn/mirtools/) This strategy was used for miRNA discovery in several protozoan and helminth parasites, for example in *A. suum* (Wang et al., 2011). As high-throughput technologies improve and with genome and transcriptome sequencing projects in progress for many parasites, it can be expected that the number of annotated novel miRNAs will soon increase. However, it is very important to perform a stringent analysis of the data obtained to confidently identify parasite miRNAs.

3.2. Identification of miRNAs in Protozoan and Helminth Parasites

Even though the explosion of miRNAs isolation and identification started in the early 2000s, it was not until 2006 that the first miRNAs from eukaryotic endoparasites of animals began to be identified (see Tables 1 and 2). Since then, many reports covering different phyla have been published. For example, de Souza Gomes et al. (2011) used homology based analysis to identify conserved candidate miRNAs in *S. mansoni*. An experimental approach based on small scale cloning was used by Cucher et al. (2011) to miRNA discovery in *E. granulosus*. The latter approach allowed the discovery of conserved as well as novel miRNAs from cestodes.

Currently, there is a fairly comprehensive view of the miRNA repertoire in parasitic protozoans and helminths. However, for further analysis we will only consider those reports that meet the following criteria:

1) Experimental isolation of the candidate miRNAs (i.e., cloning or high-throughput sequencing) or bioinformatic prediction
2) Candidate mature miRNA sequences reported
3) Candidate pre-miRNA sequences reported
4) Northern Blot validation of candidate sequences (or at least a subset) mostly when 3) cannot be performed due to the lack of a genome project of the corresponding organism or a related one (i.e., an organism of the same genus) and when no other

reports have validated by Northern Blot the presence of miRNAs in the corresponding genus.

It can be observed that post-transcriptional regulation by miRNAs seem to be a common feature throughout many protozoan parasites and for all the helminth parasites studied so far (Tables 1 and 2). Nevertheless, the presence of miRNA mechanism in one species does not imply that all the species belonging to the same phylum will present it. This is the case in Apicomplexa, where *T. gondii* and *N. caninum* have been shown to present miRNAs (Braun et al., 2010) while in *Plasmodium* spp. they are absent (Rathjen et al., 2006; Xue et al., 2008a) even though preliminary *in silico* analyses based on sequence homology search in *Plasmodium* genome with predicted *T. vaginalis* miRNAs suggested the opposite (Lin et al., 2009a). This is why we consider that the characterization of miRNA candidates should be as complete as possible in order to avoid misleading conclusions. We also encourage researchers to deposit the sequences at the miRNAs public database, miRBase (http://www.mirbase.org/), as this will facilitate comparative and identification analyses.

3.2.1. miRNAs in Protozoan Parasites

In the protozoan species *G. lamblia*, a number of miRNAs originate from snoRNA precursors (see item 2.2.1). They have been shown to depend on GlDicer for biogenesis and on GlAgo for regulation of gene expression (Saraiya and Wang, 2008; Li et al., 2011; Li et al., 2012).

Table 3. microRNAs from parasitic and free-living helminths

	Organism	N° of mature miRNAs	N° of coding genes[a]	N° of mature microRNAs/ coding genes (%)
Parasitic helminths	*Schistosoma japonicum*	78	12,669	0.6
	Schistosoma mansoni	29	12,871	0.2
	Echinococcus spp.	24	11,000	0.2
	Brugia malayi	83	11,508	0.7
	Haemonchus contortus	194	NE	-
	Ascaris suum	189	18,542	1.0
Free-living helminths	*Schmidtea mediterranea*	257	31,955	0.8
	Caenorhabditis elegans	368	19,735	1.9

a: According to Hillier et al., 2005; Ghedin et al., 2007; Robb et al., 2008; Jex et al., 2011; Olson et al., 2012; Zerlotini et al., 2013.
NE: not estimated yet.
Source: miRBase, release 19.

In the protozoan *T.gondii*, although the miRNA machinery is related to that of the unicellular algae *Chlamydomonas* (see item 2.2.1), the thermodinamical and biochemical features of precursor and mature miRNAs are characteristic from animal miRNAs (Braun et al, 2010). In addition, miRNA expression differences were observed between *T. gondii* strains with marked differences in virulence, suggesting that miRNA regulation of gene expression can influence important features such as parasite virulence (Braun et al., 2010).

So far, most protozoan miRNA sequences have no homologs among the reported sequences from other organisms (miRBase, release 19) (Saraiya and Wang, 2008; Braun et al., 2010; Saraiya et al., 2011; Li et al., 2011; Li et al., 2012).

Figure 2. Alignments of two conserved microRNAs. cel: *Caenorhabditis elegans*, hco: *Haemonchus contortus*, bma: *Brugia malayi*, asu: *Ascaris suum*, sme: *Schmidtea mediterranea*, sja: *Schistosoma japonicum*, sma: *Schistosoma mansoni*, egr: *Echinococcus granulosus*, emu: *Echinococcus multilocularis*.

3.2.2. miRNAs in Helminth Parasites

When the number of miRNAs identified in free-living and parasitic helminths is compared, taking into account only those organisms for which high scale sequencing projects have been performed, it can be observed that the value obtained is about twice in the free-living nematode *Caenorhabditis elegans* (*C. elegans*/*A. suum* = 1.9; *C. elegans*/*Haemonchus contortus* = 1.9) or thrice in the planaria *Schmidtea mediterranea* (*S. mediterranea*/*Schistosoma japonicum* = 3.3) (Table 3). On the other hand, as can be observed in Table 3, the number of discovered miRNAs in the model platyhelminth *S. mediterranea* is around 1% of the number of coding genes, while in the parasites from the genus *Schistosoma* spp. and *Echinococcus* spp. the estimated ratio is below this value. The same situation is observed among the nematodes, where *C. elegans* miRNAs genes reach ~2% of the number of coding genes and the parasitic species reach around 1% (Table 3). However, we cannot conclude that the proportion of miRNAs in helminth parasites is lower than in free-living helminths since the discovery of miRNAs in the former organisms is still incomplete and many miRNA sequences from helminth parasites have not been deposited in the miRBase. It is probable that many miRNAs are expressed in different stages that have not been analyzed yet. It is worth mentioning that for some helminth parasites, such as *E. granulosus* and *S. mansoni*, no high scale sequencing projects have been performed so far. This approach would highly increase the chances of identifying miRNAs with low expression. Another strategy when access to high throughput technologies is not available is the mix of RNA samples from different developmental stages, as was successfully performed for the nematode *Brugia malayi* (Poole et al., 2010). In the future, if the difference in miRNA number between parasitic and free living helminths is confirmed, it would be interesting to analyze how parasites perform fine-tune regulation of gene expression with a smaller miRNA repertoire.

For several helminth parasites, it was shown that miRNAs exhibit differential expression according to parasite stage. For example in *E. granulosus*, which shows a remarkable developmental plasticity, a number of conserved miRNAs were differentially expressed between cystic (metacestode) and pre-adult (protoscolex) stages, suggesting that miRNAs could have a role in the development of this parasite (Cucher et al., 2011). It is interesting to note that nematode miRNAs present more sequence conservation than in platyhelminths, with 100% identity in the sequences of conserved miRNAs such as let-7, miR-1, miR-9 and miR-71 (Figure 2). However, for all the helminth parasites studied so far novel miRNAs were also identified, and some of them were also validated. Thus, endoparasites not only show peculiarities in their miRNA machinery but also in their miRNAs repertoire.

4. What is the Job of miRNAs in Parasites?

miRNAs exert post transcriptional regulation of gene expression by guiding the effector complex miRISC to their target mRNAs. In view of this way of action, it is essential to identify the target/s of a miRNA in order to envisage its function/s. However, this is not an easy task, especially for metazoan miRNAs. Difficulties in predicting miRNA targets arise from the fact that generally animal miRNAs bind their targets by partial complementarity over a short sequence (generally 6-7 nt). The search of such a short complementary sequence in mRNAs will obviously lead to many false positive predictions. The task is further complicated by the facts that each miRNA can regulate hundreds of genes, downregulation of gene expression is often subtle, and the rules of targeting are not completely clear. For several parasites having complex life cycles that are hard to reproduce at the laboratory, more difficulties arise due to the limited availability of parasite material, lack of appropriate *in vitro* and *in vivo* systems, unavailability of annotated genomes and/or RNAseq data, lack of transfection and silencing protocols and lack of commercially available antibodies and microarrays. In addition, these parasites are cause of neglected diseases in developing countries, which implies lack of economic resources for their research. Thus, target identification is the most challenging step in miRNA research, especially in parasites. Since many options are available, it is important to decide which is the best or the possible strategy for the parasite under study. Target identification is generally performed by a combination of bioinformatic prediction and experimental strategies (Figure 3). Here, we summarize computational and experimental strategies to identify miRNA targets.

4.1. Bioinformatic Prediction of miRNA Targets

Although many miRNA targets have been computationally predicted, only a limited number of them were supported experimentally. The biological function of most miRNAs therefore remains unknown. In metazoans, miRNAs appear to primarily target the 3' untranslated region (UTR) of the mRNA; but miRNA recognition elements in the coding sequence or 5'untranslated region have also been reported (Thomas et al., 2010 and references therein). Target prediction relies on: (i) complementarity between 3'UTRs and the 5' seed region of the miRNA (position 2-8); (ii) thermodynamics of the miRNA-mRNA duplex; (iii) conservation of binding sites across orthologous 3'UTRs. However, there are algorithms that also consider target recognition sites outside the seed of the miRNA and in the ORF and 5'UTR (Thomas et al., 2010 and references therein). Thus, to identify miRNA targets in a given species, it is important to get reliable 3'UTR and mature miRNA sequences since most algorithms for miRNA target prediction requires both of them as input datasets. In spite that genome and transcriptome sequencing projects are now available for a whole range of parasites, 3'UTRs have not been characterized and annotated for many of them. However, they could be determined by selecting a downstream flanking sequence of the stop codon of the genes since databases like GeneDB (http://www.genedb.org) provides reliable access to genome sequences and annotation data for these pathogens and closely related organisms.

4.2. Algorithms for miRNA Target Prediction

Some of the most common algorithms for miRNA target prediction are miRanda (Enright et al., 2003), TargetScan (Lewis et al., 2003), PicTar (Krek et al., 2005), RNAhybrid (Rehmsmeier et al., 2004) and rna22 (Miranda et al., 2006). It remains difficult to determine which of the methods mentioned above is more appropriate to predict miRNA targets in parasites since all algorithms were able to identify previously predicted or experimentally verified targets of model organisms for which they were used, and were not trained or verified with parasite miRNA targets. To compare the efficiency of the methods, the same dataset of validated miRNA-targets predictions should be assesed by different programs, but this is not possible to be performed in parasites since very few targets were experimentally validated among protozoan parasites and no one for parasitic helminths. This suggests that efforts should be made to experimentally validate targets for parasite miRNAs. With genome and transcriptome projects ongoing for many parasites, it is expected an improvement in the annotated sequence data that will enable better predictions. Additionally, with the unraveling of specific features of miRNA targets in parasites, novel and accurate algorithms suitable for these organisms could be developed. Nevertheless, experimental validation is essential to identify *bona fide* miRNA targets since false positive predictions from the programs are expected.

4.3. Experimental Identification of miRNA Targets

One possible start-point is to analyze the mRNA and/or protein expression profiles by high-throughput mRNA sequencing or proteomics respectively, combined with miRNA profiling in different developmental stages or conditions. Those genes whose expression levels are inversely correlated with the miRNA concentration, can be considered candidate miRNA targets. Also, low protein levels combined with high mRNA expression of a particular gene can suggest miRNA regulation at the translational level. This last consideration was taken into account for miRNA target identification in *T. vaginalis* (see next section). Large-scale miRNA target identification can be performed in order to have a landscape of targets regulated by a single or a small number of miRNAs, or to know which mRNAs are being subjected to miRNA regulation in a cell, tissue or organism. For the first approach differential expression of a single miRNA is followed by mRNA or protein expression global analyses and for the second strategy, immunoprecipitation of mRNAs associated to Ago proteins or other miRISC components is generally followed by high throughput sequencing of immunoprecipitated mRNAs (Figure 3). Next, a brief description of these strategies and their applications to parasites follows.

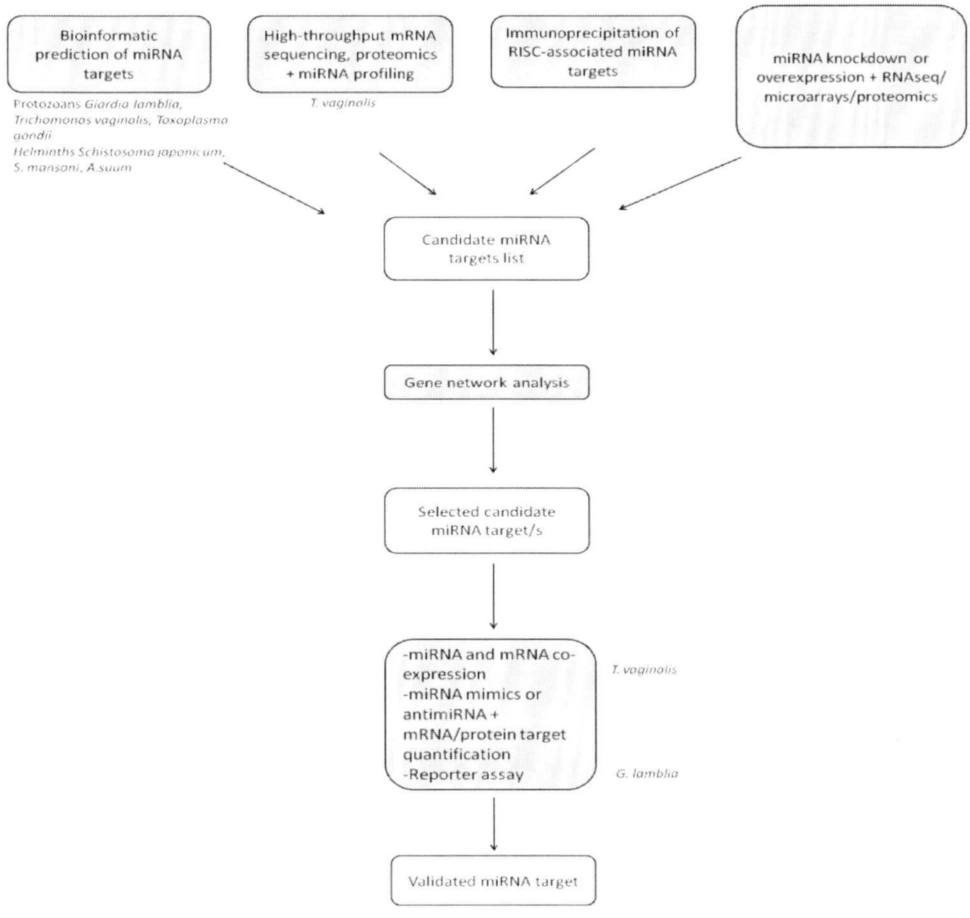

Figure 3. Methods for miRNA target identification. Putative target genes can be identified by bioinformatic approaches, analyses of protein, mRNA and miRNA expression profiles, overexpression or knockdown of miRNA followed by expression profiling or immunoprecipitation of miRNA targets associated to miRISC. The candidate miRNA targets list obtained by these methods (generally hundreds of targets) can be reduced by selection according to Gene Network analysis: Gene Ontology (GO) terms, enriched biological pathways or gene interaction networks. Selected targets can be validated by checking that the miRNA and its target are co-expressed, overexpression or knockdown of the miRNA followed by RNA or protein target quantification or by the reporter assay. The parasite species for which these methods were applied are indicated outside the boxes.

I. **miRNA overexpression:** overexpression of a miRNA can be achieved by transient transfection of a specific pre-miRNA (Johnson et al., 2007) or a synthetic dsRNA designed to mimic mature endogenous miRNA (Vinther et al., 2006). Another possible strategy is to overexpress a miRNA by transfection with biotynilated synthetic miRNA and to capture the miRNA-mRNA complexes with streptavidin beads followed by purification and analysis of mRNAs (Orom & Lund, 2007). The mRNA analyses can be done in a large scale, for example by microarrays (Lal et al., 2011) or the enrichment of a particular candidate miRNA target with respect to a proper control can be analyzed for example by RT-qPCR.

II. **miRNA knockdown:** the inhibition of miRNAs can be achieved by transfection of antisense oligonucleotides (ASOs) that block miRNA function probably by binding mature miRNAs in miRISCs (Horwich & Zamore, 2008) competing with target mRNA binding. A variety of chemically modified ASOs have been developed in order to make them resistant to unspecific cellular ribonucleases and to promote a high affinity binding to miRNAs so as to effectively out-compete binding to target mRNA. Chemically modified ASOs include 2'-O-methyl oligonucleotides (Leaman et al, 2005; Horwich and Zamore, 2008), Locked Nucleic Acid (LNA) nucleotides (Elmen et al., 2008a; Elmen et al., 2008b), cholesterol conjugated oligonucleotides (Krutzfeldt et al., 2005; Horwich & Zamore, 2008 and references therein) and morpholinos (Kloosterman et al., 2007). An alternative for miRNA function inhibition are the "miRNA sponges". They specifically inhibit microRNAs with a complementary heptameric seed, such that a single sponge can be used to block an entire microRNA seed family (Ebert et al., 2007).

For some model organisms such as human and mouse it was observed that miRNAs reduce the transcript levels of most target genes (Thomas et al., 2010). Thus, detecting downregulated mRNAs after exogenous expression of a miRNA or upregulated mRNAs after miRNA knockdown may allow identifying miRNA targets. However, it is unknown if the same occurs in protozoan or helminth parasites, the few examples of miRNA overexpression in parasites (see below) show that the RNA target level is not altered. Another possibility is to analyze changes in protein profile by proteomics as described in Selbach et al. (2008).

One important limitation of all these silencing and overexpression techniques is that they can only be applied to parasites in which transfection and silencing protocols are very well set up.

III. **Methods based on immunoprecipitation of miRISC proteins:** an alternative approach has been described to miRNA target identification based on stable *in vivo* interaction between the miRISC and the target mRNA. Antibodie*s again*st proteins of the miRISC complex, such as Ago, are used for immunoprecipitation and the co-precipitated mRNA are analyzed by microarray and/or high throughput RNAseq. Although this biochemical approach has not been reported in parasites, results in model organisms (Zhang et al., 2007) demonstrate that this is an effective procedure to identify miRNA targets that could be used in parasites since miRISC components have been identified and characterized in some of them. For example, Cai et al. (2012) applied this approach to identify and char*a*cterize endogenous Ago2 and its associated small RNAs in *S. japonicum* highlighting the potential of the method to *be applied t*o identify its associated miRNA targets. One potential drawback is that it may be necessary a large amount of starting material and many parasites are difficult to obtain or cultivate. It must be taken into account that this method captures a subset of stable miRNA targets associations (leading to the loss of targets) and does not identify mRNAs associated with a specific miRNA unless a particular miRNA is overexpreesed or knocked out prior to inmunoprecipitation.

Another approach that improves *in vivo* target identification is crosslinking inmunoprecipitation followed by deep sequencing (HITS-CLIP or CLIP-seq). This technique uses UV-irradiation to covalently crosslink RNA to miRISC proteins that are then inmunoprecipitated and coupled with high-throughput sequencing to identify target sites of bound RNAs (Chi et al., 2009).

4.3.1. Gene Network Analysis

Since one miRNA can regulate many genes, it is important to identify key miRNA targets and functions among the many target candidates. The molecular function and biological process can be analyzed by the program Blast2GO using the GO tool CateGOrizer (www.animalgenome.org /bioinfo/tools/countgo/). In addition, the analysis of interactions among the candidate target genes by softwares such as Ingenuity Pathways (Thomas & Bonchev, 2010) can allow to determine which genes have highly connected nodes which are predicted to be biologically important miRNA targets (Thomas et al., 2010).

4.3.2. Validation of miRNA Targets

After identification of putative miRNA targets by large scale approaches, and reducing the list by gene pathway and interaction analyses, it is necessary to validate candidate miRNA targets (Figure 3). The following strategies can be applied to this end:

I. **miRNA and target mRNA co-expression:** very often the list of candidate targets is just the result of bioinformatic analyses and it is unknown whether the expression of a miRNA and its putative target are coincident which is obviously necessary for interaction. Co-expression can be analyzed using total RNA extracted from the cells or tissue under study and performing Northern blot technique for mRNA and for small RNAs, or by RT- qPCR for mRNA and stem-loop RT-qPCR (Chen et al., 2005) for detecting the miRNA. Also, spatial co-expression can be analyzed by *in situ* hybridization, see for example Johnson et al. (2007).

II. **Reporter assay**: this validation assay can be performed by cloning the 3'UTR of the putative target gene downstream of the ORF of a reporter gene such as green fluorescent protein (GFP) or luciferase. The recombinant plasmid and the miRNA of interest are then transfected in a cell which preferably has low or null expression of the miRNA under analysis. At the corresponding time after transfection, fluorescence or luciferase activity are determined. The expected result for a true target-miRNA interaction is a reduction of these activities (Kuhn et al., 2008). This assay is generally performed in a heterologous system, but it can also be performed in the organism/cell under study, as it was performed for *Giardia* (see below). The reporter assay allows assessing direct regulation of gene expression by a miRNA.

III. **miRNA overexpression or knockdown followed by analysis of selected target expression**: an inverse relationship between the gene product (at least the protein) and the miRNA is expected. The target mRNA expression can be evaluated by RT-qPCR or Northern blot. The protein level can be analyzed by Western blot if antibodies are available; other possibilities such as two dimensional gel electrophoresis and MALDI-TOF analysis are also possible, as it was performed for *T. vaginalis* miRNA target identification (see below).

4.4. miRNA Targets in Endoparasites

Although miRNAs were reported for a wide range of protozoan and helminth parasites (see section 3), miRNA targets were identified only for a few of them (see figure 3).

A search for target mRNAs was performed for a small number of *G. lamblia* miRNAs by a combination of computational and experimental strategies. The prediction programs used were miRanda and RNAhybrid, which rendered 95% coincident results. Since 3'UTRs in *Giardia* are very short, segments including 50-100 nts upstream and 50-200 nts downstream from the stop codon of each ORF were considered. A proportion of the 100-200 candidate targets obtained for snoRNA derived as well as ORF derived miRNAs were Variant Surface Proteins (VSPs). Since the expression of a single VSP at *Giardia* surface is related to evasion of the host immune response, VSPs were chosen for further experimental analysis. The reporter assay was employed for miRNA target validation: *in vitro* transcribed RNA containing a number of copies of the putative VSP target site in the *Renilla* luciferase gene 3' UTR was electroporated into *Giardia* WB trophozoites together with the chemically synthesized miRNA and the luciferase activity was measured after proper incubation and cell lysis. Small but significant reduction of luciferase activity was observed, suggesting that the VSPs could be targeted by miRNAs. Since no change in mRNA levels was detected it was inferred that the tested miRNAs acted by repression of translation. The reduction of luciferase activity after miRNA overexpression was not observed in *Giardia* cells knocked down for GIAgo. In addition, miRNA knockdown by the co-transfection with a 2'-O-methyl oligonucleotide complementary to the miRNA under evaluation resulted in recovery of luciferase activity. In other assay, miRNA knockdown was performed in cells transfected with the reporter, resulting in an increase of luciferase activity, which suggests that the endogenous miRNA is repressing the expression of the reporter gene containing the VSP target sites. Additional strategies included overexpression of miRNAs in *Giardia* cells transfected with a c-myc tagged VSP followed by Western blot examination of VSP expression level using an antibody against c-myc. The results obtained suggested involvement of a miRNA-mediated mechanism in controlling VSP expression (Saraiya & Wang 2008; Saraiya et al., 2011; Li et al., 2011; Li et al., 2012). However, other possible mechanism based on RNAi was proposed for the control of VSPs expression in *Giardia* (Prucca et al., 2008).

In *T. vaginalis*, the search for potential miRNA targets was performed by first identifying genes with high mRNA expression but low protein level whose translation could be potentially regulated by miRNAs. Because of the lack of commercially available antibodies against proteins from protozoan parasites, the analysis of protein expression was performed by two dimensional gel electrophoresis followed by MALDI-TOF analysis. Malate dehydrogenase (MDH) was selected as one of the candidate miRNA targets and further analyzed. Potential miRNA binding sites were searched by RNA22 program (Miranda et al., 2006) using nine *T. vaginalis* miRNAs previously identified (Lin et al., 2009b) and the miRNA sequences available at miRBase. Four binding sites for the highly conserved miR-1 family members were predicted, three were localized in the open reading frame while one was in the 3'UTR. After confirming the presence of miR-1 gene in the parasite genome and its expression in the stage that showed low MDH protein level, miRNA overexpression was performed with commercially available human miR-1 mimic transfected into low miR-1 expressing trophozoite *T. vaginalis* cells. As a result the MDH protein expression was significantly lowered in comparison to a control, suggesting that MDH is targeted by miR-1 in *T. vaginalis*.

Bioinformatic search was performed in order to find miRNA targets in *T. gondii*, a parasite that shows peculiarities in its miRNA machinery (see section 2.2.1). Most of the 14 miRNAs identified have target sites in a variety of transcripts. Interestingly, targets exhibit perfect or near perfect complementarity to miRNAs, a characteristic that is very rarely observed in metazoan miRNAs. Most miRNAs analyzed have target sites in the 3'UTR of coding genes, but also in the 5' UTRs and open reading frames. It would be interesting to experimentally validate these miRNA targets, since the predicted interactions with miRNAs show unique characteristics: they are perfect or near perfect interactions that would be exerted by miRNAs associated to slicer independent Ago that seems to regulate target expression at the translational level.

The search for miRNA targets in helminth parasites was performed only by bioinformatic methods. The softwares used were miRanda and RNAhybrid. To identify which biological processes, molecular functions or cellular components may be targeted by the miRNAs previously identified, the predicted miRNA gene targets were categorized based on their Gene Ontology (GO) terms using the GO tool CateGOrizer (www.animalgenome.org /bioinfo/tools/countgo/). This classification method allowed the authors to count and group the GO classes of the predicted targets and identify which are more represented in the set of miRNA gene targets (de Souza Gomes et al., 2011; Xu et al., 2013). For functional information about the targets genes KEGG (Kyoto Encyclopedia of Genes and Genomes) or Blast2GO were used.

Potential target genes for 176 *S. japonicum* miRNAs were predicted using miRanda software, via searching against the 3´UTR of *S. japonicum* protein-coding genes (Huang et al., 2009). The 3'UTRs were predicted by integrating the predicted mRNAs from genome data and EST information. The results showed that each miRNA could regulate many target genes, from several to hundreds, and some development-associated genes could be regulated by many miRNAs. For example, TGF-β receptor II/activin receptor IIA, which plays important roles in schistosomal development including male-induced female reproductive development and host-parasite interaction, could be regulated by 16 miRNAs, including let-7, a miRNA known to regulate developmental timing in *C. elegans*. Some members of the cathepsin superfamily essential to the mammal-parasitic stages of schistosomes, could be regulated by several miRNAs. Some genes, such as insulin-induced gene 1 protein (INSIG-1), cytosolic thyroid hormone binding protein and thyroid hormone receptor interactor 10, possibly associated with the exploitation of host hormones, could be modulated by *S. japonicum* miR-31 and miR-2d (Huang et al., 2009).

In *S. mansoni* miRNA targets were also predicted using the miRanda software. De Souza Gomes et al. (2011) identified 389 mRNA potential targets for the 67 mature miRNAs previously identified by bioinformatic methods. The majority of the 3′ UTRs chosen for analysis in this study were also represented by *S. mansoni* ESTs to improve the confidence of the findings. Of the 67 mature *S. mansoni* miRNAs identified, 50 of these miRNAs had at least one putative target *S. mansoni* gene. A major finding was that the conserved miRNA sma-miR-71 had the highest number of targets compared to others as it was predicted to potentially regulate 42 distinct target genes. The clusters sma-mir-71/2 and sma-mir-71b/2, which contain together in total 12 different mature miRNAs, combined have 200 predicted targets, meaning that more than 50% of all the predicted targets could be regulated by these two miRNAs. A number of the *S. mansoni* miRNA gene targets can be targeted by more than one mature miRNA (e.g. sma-miR-71b-3p and sma-miR-31-5p that target the Smp_184330 hypothetical gene). Moreover, a number of miRNAs have multiple target sites within the same target gene (e.g. sma-miR-2b-5p that contains target sites in different positions in the 3′ UTR of the hypothetical gene Smp_110400.3). The Gene Ontology analysis showed that catalytic activity and metabolism related GO terms represented almost 20% of the GO terms of *S. mansoni* miRNA target genes analyzed. Using KEGG, the term metabolic pathway was over represented showing 14 of the targets analyzed.

For target prediction of the nematode *A. suum* (Xu et al., 2013), all mRNA and ESTs of *A. suum* deposited in the NCBI were used including the gender-enriched ESTs. The predicted numbers of targets per miRNA ranged from 1 to thousands for gene-specific as well as shared miRNAs. Functional prediction revealed a set of elongation factors, heat shock proteins and growth factors from the targets of gender-specific miRNAs, which suggested that the miRNAs play essential functions in parasite development. Other important genes targeted by gender-specific miRNAs are MSPs, which are proteins involved in sperm motility.

It is known that miRNAs may act in concert to regulate target genes, or may have multiple target sites within the same target gene as well as regulate many different transcripts. The results of these studies showed that this also happens among helminth parasites, suggesting that miRNAs could be involved in the control of essential parasite pathways. It would be interesting to experimentally validate some of the predicted miRNA targets to further elucidate their role in parasite biology.

Conclusion

miRNA regulation of gene expression is starting to be unraveled in protozoan and helminth parasites. So far, it has been shown that miRNA machinery and miRNA repertoire from endoparasites have specific characteristics, some of them unshared with their hosts. For many protozoan and all helminth parasites analyzed so far, miRNA expression patterns and abundance suggest important roles for parasite biology. Although further research is necessary to know about miRNA targeting in parasites, some of the studies that identified targeted genes, suggested involvement of miRNAs in essential processes such as evasion of host immune response. Expected advances in genomic projects and functional genomics for these organisms will lead to a deeper knowledge on miRNAs and the pathways they regulate. This will in turn provide new diagnosis and intervention strategies for the devastating diseases they produce.

References

Ai, L.; Xu, M.J.; Chen, M.X.; Zhang, Y.N.; Chen, S.H.; Guo, J.; Cai, Y.C.; Zhou, X.N.; Zhu, X.Q.; Chen, J.X. (2012). Characterization of microRNAs in *Taenia saginata* of zoonotic significance by Solexa deep sequencing and bioinformatics analysis. *Parasitology Research, 110*, 2373-2378.

Al Riyahi, A.; Al-Anouti, F.; Al-Rayes, M.; Ananvoranich, S. (2006). Single argonaute protein from *Toxoplasma gondii* is involved in the double-stranded RNA induced gene silencing. *International Journal for Parasitology, 36*, 1003-1014.

Artzi, S.; Kiezun, A.; Shomron, N. (2008). miRNAminer: a tool for homologous microRNA gene search. *BMC Bioinformatics, 9*, 39.

Babiarz, J.E.; Ruby, J.G.; Wang, Y.; Bartel, D.P.; Blelloch, R. (2008). Mouse ES cells express endogenous shRNAs, siRNAs, and other Micro-processor-independent, Dicer-dependent small RNAs. *Genes & Development, 22*, 2773–2785.

Bartel, D.P. (2004). MicroRNAs: genomics, biogenesis, mechanism, and function. *Cell, 116*, 281-297.

Bogerd, H.P.; Karnowski, H.W.; Cai, X.; Shin, J.; Pohlers, M.; Cullen, B.R. (2010). A mammalian herpesvirus uses noncanonical expresión and processing mechanisms to generate viral microRNAs. *Molecular Cell, 37*, 135–42

Braun, L.; Cannella, D.; Ortet, P.; Barakat, M.; Sautel, C.F.; Kieffer, S.; Garin, J.; Bastien, O.; Voinnet, O.; Hakimi, M.A. (2010). A Complex Small RNA Repertoire Is Generated by a Plant/Fungal-Like Machinery and Effected by a Metazoan-Like Argonaute in the Single-Cell Human Parasite *Toxoplasma gondii. PLoS Pathogens, 6*, e1000920.

Cai, P.; Hou, N.; Piao, X.; Liu, S.; Liu, H.; Yang, F.; Wang, J.; Jin, Q.; Wang, H.; Chen, Q. (2011). Profiles of Small Non-Coding RNAs in *Schistosoma japonicum* during Development. *PLoS Neglected Tropical Diseases, 5*, e1256.

Cai, P.; Piao, X.; Hou, N.; Liu, S.; Wang, H.; Chen, Q. (2012). Identification and Characterization of Argonaute Protein, Ago2 and Its Associated Small RNAs in *Schistosoma japonicum. PLoS Neglected Tropical Diseases, 6*, e1745.

Chen, C.; Ridzon, D.A.; Broomer, A.J.; Zhou, Z.; Lee, D.H.; Nguyen, J.T.; Barbisin, M.; Xu, N.L.; Mahuvakar, V.R.; Andersen, M.R.; Lao, K.Q.; Livak, K.J.; Guegler, K.J. (2005). Real-time quantification of microRNAs by stem-loop RT-PCR. *Nucleic Acids Research, 33*, e179.

Chen, X.S.; Collins, L.J.; Biggs, P.J.; Penny, D. (2009). High Throughput Genome-Wide Survey of Small RNAs from the Parasitic Protists *Giardia intestinalis* and *Trichomonas vaginalis. Genome Biology and Evolution, 1*, 165-175.

Chen, M.X.; Ai, L.; Xu, M.J.; Chen, S.H.; Zhang, Y.N.; Guo, J.; Cai, Y.C.; Tian, L.G.; Zhang, L.L.; Zhu, X.Q.; Chen, J.X. (2011a). Identification and characterization of microRNAs in *Trichinella spiralis* by comparison with *Brugia malayi* and *Caenorhabditis elegans. Parasitology Research, 109*, 553–558.

Chen, M.X.; Ai, L.; Xu, M.J.; Zhang, R.L.; Chen, S.H.; Zhang, Y.N.; Guo, J.; Cai, Y.C.; Tian, L.G.; Zhang, L.L.; Zhu, X.Q.; Chen, J.X. (2011b). *Angiostrongylus cantonensis*: Identification and characterization of microRNAs in male and female adults. *Experimental Parasitology, 128*, 116-120.

Chi, S.W.; Zang, J.B.; Mele, A.; Darnell, R.B. (2009). Argonaute HITS-CLIP decodes microRNA-mRNA interaction maps. *Nature*, *460*, 479-486.

Copeland, C.S.; Marz, M.; Rose, D.; Hertel, J.; Brindley, P.J.; Santana, C.B.; Kehr, S.; Attolini, C.S.; Stadler, P.F. (2009). Homology-based annotation of non-coding RNAs in the genomes of *Schistosoma mansoni* and *Schistosoma japonicum*. *BMC Genomics*, *10*, 464.

Cucher, M.; Prada, L.; Mourglia-Ettlin, G.; Dematteis, S.; Camicia, F.; Asurmendi, S.; Rosenzvit, M. (2011). Identification of *Echinococcus granulosus* microRNAs and their expresión in different life cycle stages and parasite genotypes. *International Journal for Parasitology*, *41*, 439–448.

De, D.; Pal, D.; Ghosh, S.K. (2006). *Entamoeba histolytica*: Computational identification of putative microRNA candidates. *Experimental Parasitology*, *113*, 239–243.

de Souza Gomes, M.; Muniyappa, M.K.; Carvalho, S.G.; Guerra-Sá, R.; Spillane, C. (2011). Genome-wide identification of novel microRNAs and their target genes in the human parasite *Schistosoma mansoni*. *Genomics*, *98*, 96-111.

Ebert, M.S.; Neilson, J.R.; Sharp, P.A. (2007). MicroRNA sponges: competitive inhibitors of small RNAs in mammalian cells. *Nature Methods*, *4*, 721-726.

Elmen, J.; Lindow, M.; Silahtaroglu, A.; Bak, M.; Christensen, M.; Lind-Thomsen, A.; Hedtjärn, M.; Hansen, J.B.; Hansen, H.F.; Straarup, E.M.; McCullagh, K.; Kearney, P.; Kauppinen, S. (2008a). Antagonism of microRNA-122 in mice by systemically administered LNA-antimiR leads to up-regulation of a large set of predicted target mRNAs in the liver. *Nucleic Acids Research*, *36*, 1153-1162.

Elmen, J.; Lindow, M.; Schütz, S.; Lawrence, M.; Petri, A.; Obad, S.; Lindholm, M.; Hedtjärn, M.; Hansen, H.F.; Berger, U.; Gullans, S.; Kearney, P.; Sarnow, P.; Straarup, E.M.; Kauppinen, S. (2008b). LNA-mediated microRNA silencing in non-human primates. *Nature*, *452*, 896-899.

Enright, A.J.; John, B.; Gaul, U.; Tuschl, T.; Sander, C.; Marks, D.S. (2003). MicroRNA targets in *Drosophila*. *Genome Biology*, *5*, R1.

Gajria, B.; Bahl, A.; Brestelli, J.; Dommer, J.; Fischer, S.; Gao, X.; Heiges, M.; Iodice, J.; Kissinger, J.C.; Mackey, A.J.; Pinney, D.F.; Roos, D.S.; Stoeckert, C.J. Jr.; Wang, H.; Brunk, B.P. (2008). ToxoDB: an integrated *Toxoplasma gondii* database resource. *Nucleic Acids Research*, *36*, D553-D556.

Ghedin, E.; Wang, S.; Spiro, D.; Caler, E.; Zhao, Q.; Crabtree, J.; Allen, J.E.; Delcher, A.L.; Guiliano, D.B.; Miranda-Saavedra, D.; Angiuoli, S.V.; Creasy, T.; Amedeo, P.; Haas, B.; El-Sayed, N.M.; Wortman, J.R.; Feldblyum, T.; Tallon, L.; Schatz, M.; Shumway, M.; Koo, H.; Salzberg, S.L.; Schobel, S.; Pertea, M.; Pop, M.; White, O.; Barton, G.J.; Carlow, C.K.; Crawford, M.J.; Daub, J.; Dimmic, M.W.; Estes, C.F.; Foster, J.M.; Ganatra, M.; Gregory, W.F.; Johnson, N.M.; Jin, J.; Komuniecki, R.; Korf, I.; Kumar, S.; Laney, S.; Li, B.W.; Li, W.; Lindblom, T.H.; Lustigman, S.; Ma, D.; Maina, C.V.; Martin, D.M.; McCarter, J.P.; McReynolds, L.; Mitreva, M.; Nutman, T.B.; Parkinson, J.; Peregrín-Alvarez, J.M.; Poole, C.; Ren, Q.; Saunders, L.; Sluder, A.E.; Smith, K.; Stanke, M.; Unnasch, T.R.; Ware, J.; Wei, A.D.; Weil, G.; Williams, D.J.; Zhang, Y.; Williams, S.A.; Fraser-Liggett, C.; Slatko, B.; Blaxter, M.L.; Scott, A.L. (2007). Draft genome of the filarial nematode parasite *Brugia malayi*. *Science*, *317*, 1756-1760.

Ghildiyal, M. & Zamore, P.D. (2009). Small silencing RNAs: an expanding universe. *Nature Reviews Genetics*, *10*, 94-108.

Hao, L.; Cai, P.; Jiang, N.; Wang, H.; Chen, Q. (2010). Identification and characterization of microRNAs and endogenous siRNAs in *Schistosoma japonicum*. *BMC Genomics*, *11*, 55.

Hillier, L.W.; Coulson, A.; Murray, J.I.; Bao, Z.; Sulston, J.E.; Waterston, R.H. (2005). Genomics in *C. elegans*: so many genes, such a little worm. *Genome Research*, *15*, 1651-1660.

Hofacker, I.L. (2003). Vienna RNA secondary structure server. *Nucleic Acids Research*, *31*, 3429-3431.

Horwich, M.D. & Zamore, P.D. (2008). Design and delivery of antisense oligonucleotides to block microRNA function in cultured *Drosophila* and human cells. *Nature Protocols*, *3*, 1537-1549.

Huang, J.; Hao, P.; Chen, H.; Hu, W.; Yan, Q.; Liu, F.; Han, Z.G. (2009). Genome-Wide Identification of *Schistosoma japonicum* MicroRNAs Using a Deep-Sequencing Approach. *PLoS ONE*, *4*, e8206.

Huang, P.J.; Lin, W.C.; Chen, S.C.; Lin, Y.H.; Sun, C.H.; Lyu, P.C.; Tang, P. (2012). Identification of putative miRNAs from the deep-branching unicellular flagellates. *Genomics*, *99*, 101-107.

Jenkins, S.J. & Allen, J.E. (2010). Similarity and diversity in macrophage activation by nematodes, trematodes, and cestodes. *Journal of Biomedicine and Biotechnology*, *2010*, 262609.

Jex, A.R.; Liu, S.; Li, B.; Young, N.D.; Hall, R.S.; Li, Y.; Yang, L.; Zeng, N.; Xu, X.; Xiong, Z.; Chen, F.; Wu, X.; Zhang, G.; Fang, X.; Kang, Y.; Anderson, G.A.; Harris, T.W.; Campbell, B.E.; Vlaminck, J.; Wang, T.; Cantacessi, C.; Schwarz, E.M.; Ranganathan, S.; Geldhof, P.; Nejsum, P.; Sternberg, P.W.; Yang, H.; Wang, J.; Wang, J.; Gasser, R.B. (2011). *Ascaris suum* draft genome. *Nature*, *479*, 529-533.

Johnson, C.D.; Esquela-Kerscher, A.; Stefani, G.; Byrom, M.; Kelnar, K.; Ovcharenko, D.; Wilson, M.; Wang, X.; Shelton, J.; Shingara, J.; Chin, L.; Brown, D.; Slack, F.J. (2007). The let-7 microRNA represses cell proliferation pathways in human cells. *Cancer Research*, *67*, 7713-7722.

Kloosterman, W.P.; Lagendijk, A.K.; Ketting, R.F.; Moulton, J.D.; Plasterk, R.H. (2007). Targeted inhibition of miRNA maturation with morpholinos reveals a role for miR-375 in pancreatic islet development. *PLoS Biology*, *5*, e203.

Krautz-Peterson, G. & Skelly, P.J. (2008). *Schistosoma mansoni*: the dicer gene and its expression. *Experimental Parasitology*, *118*, 122-128.

Krek, A.; Grün, D.; Poy, M.N.; Wolf, R.; Rosenberg, L.; Epstein, E.J.; MacMenamin, P.; da Piedade, I.; Gunsalus, K.C.; Stoffel, M.; Rajewsky, N. (2005). Combinatorial microRNA target predictions. *Nature Genetics*, *37*, 495-500.

Krützfeldt, J.; Rajewsky, N.; Braich, R.; Rajeev, K.G.; Tuschl, T.; Manoharan, M.; Stoffel, M. (2005). Silencing of microRNAs in vivo with 'antagomirs'. *Nature*, *438*, 685-689.

Kuhn, D.E.; Martin, M.M.; Feldman, D.S.; Terry, A.V.Jr.; Nuovo, G.J.; Elton, T.S. (2008). Experimental validation of miRNA targets. *Methods*, *44*, 47–54.

Lagos-Quintana, M.; Rauhut, R.; Lendeckel, W.; Tuschl, T. (2001). Identification of novel genes coding for small expressed RNAs. *Science*, 294, 853-858.

Lal, A.; Thomas, M.P.; Altschuler, G.; Navarro, F.; O'Day, E.; Li, X.L.; Concepcion, C.; Han, Y.C.; Thiery, J.; Rajani, D.K.; Deutsch, A.; Hofmann, O.; Ventura, A.; Hide, W.; Lieberman, J. (2011). Capture of microRNA-bound mRNAs identifies the tumor suppressor miR-34a as a regulator of growth factor signaling. *PLoS Genetics*, *7*, e1002363.

Lau, N.C.; Lim, L.P.; Weinstein, E.G.; Bartel, D.P. (2001). An abundant class of tiny RNAs with probable regulatory roles in *Caenorhabditis elegans*. *Science*, 294, 858–862.

Leaman, D.; Chen, P.Y.; Fak, J.; Yalcin, A.; Pearce, M.; Unnerstall, U; Marks, D.S.; Sander, C.; Tuschl, T.; Gaul, U. (2005). Antisense-mediated depletion reveals essential and specific functions of microRNAs in *Drosophila* development. *Cell*. *121*, 1097-1108.

Lewis, B.P.; Shih, I.H.; Jones-Rhoades, M.W.; Bartel, D.P.; Burge, C.B. (2003). Prediction of mammalian microRNA targets. *Cell*, *115*, 787-798.

Li, W.; Saraiya, A.A.; Wang, C.C. (2011). Gene regulation in *Giardia lambia* involves a putative microRNA derived from a small nucleolar RNA. *PLoS Neglected Tropical Diseases*, *5*, e1338.

Li, W.; Saraiya, A.A.; Wang, C.C. (2012). The profile of snoRNA-derived microRNAs that regulate expression of variant surface proteins in *Giardia lamblia*. *Cellular Microbiology*, *14*, 1455–1473.

Lim, L.P.; Lau, N.C.; Garrett-Engele, P.; Grimson, A.; Schelter, J.M.; Castle, J.; Bartel, D.P.; Linsley, P.S.; Johnson, J.M. (2005). Microarray analysis shows that some microRNAs downregulate large numbers of target mRNAs. *Nature*, *433*, 769-773.

Lin, W.C.; Li, S.C.; Lin, W.C.; Shin, J.W.; Hu, S.N.; Yu, X.M.; Huang, T.Y.; Chen, S.C.; Chen, H.C.; Chen, S.J.; Huang, P.J.; Gan, R.R.; Chiu, C.H.; Tang, P. (2009a). Identification of microRNA in the protist *Trichomonas vaginalis*. *Genomics*, *93*, 487–493.

Lin, W.C.; Huang, K.Y.; Chen, S.C.; Huang, T.Y.; Chen, S.J.; Huang, P.J.; Tang, P. (2009b). Malate dehydrogenase is negatively regulated by miR-1 in *Trichomonas vaginalis*. *Parasitology Research*, *105*, 1683–1689.

Liu, F.; Lu, J.; Hu, W.; Wang, S.Y.; Cui, S.J.; Chi, M.; Yan, Q.; Wang, X.R.; Song, H.D.; Xu, X.N.; Wang, J.J.; Zhang, X.L.; Zhang, X.; Wang, Z.Q.; Xue, C.L.; Brindley, P.J.; McManus, D.P.; Yang, P.Y.; Feng, Z.; Chen, Z.; Han, Z.G. (2006). New perspectives on host-parasite interplay by comparative transcriptomic and proteomic analyses of *Schistosoma japonicum*. *PLoS Pathogens*, *2*, e29.

Liu, X.; Song, Y.; Lu, H.; Tang, B.; Piao, X.; Hou, N.; Peng, S.; Jiang, N.; Yin, J.; Liu, M.; Chen, Q. (2011). Transcriptome of Small Regulatory RNAs in the Development of the Zoonotic Parasite *Trichinella spiralis*. *PLoS ONE*, *6*, e26448.

Macrae, I.J.; Zhou, K.; Li, F.; Repic, A.; Brooks, A.N.; Cande, W.Z.; Adams, P.D.; Doudna, J.A. (2006). Structural basis for double-stranded RNA processing by Dicer. *Science*, *311*, 195-198.

Mallick, B.; Ghosh, Z.; Chakrabarti, J. (2008). MicroRNA switches in *Trypanosoma brucei*. *Biochemical and Biophysical Research Communications*, *372*, 459-463.

Militello, K.T.; Refour, P.; Comeaux, C.A.; Duraisingh, M.T. (2008). Antisense RNA and RNAi in protozoan parasites: working hard or hardly working? *Molecular and Biochemical Parasitology*, *157*, 117-126.

Miranda, K.C.; Huynh, T.; Tay, Y.; Ang, Y.S.; Tam, W.L.; Thomson, A.M.; Lim, B.; Rigoutsos, I. (2006). A pattern-based method for the identification of MicroRNA binding sites and their corresponding heteroduplexes. *Cell 126*, 1203–1217.

Miyoshi, K.; Miyoshi, T.; Siomi, H. (2010). Many ways to generate microRNA-like small RNAs: non-canonical pathways for microRNA production. *Molecular Genetics and Genomics*, *284*, 95-103.

Morrison, H.G.; McArthur, A.G.; Gillin, F.D.; Aley, S.B.; Adam, R.D.; Olsen, G.J.; Best, A.A.; Cande, W.Z.; Chen, F.; Cipriano, M.J.; Davids, B.J.; Dawson, S.C.; Elmendorf, H.G.; Hehl, A.B.; Holder, M.E.; Huse, S.M.; Kim, U.U.; Lasek-Nesselquist, E.; Manning, G.; Nigam, A.; Nixon, J.E.; Palm, D.; Passamaneck, N.E.; Prabhu, A.; Reich, C.I.; Reiner, D.S.; Samuelson, J.; Svard, S.G.; Sogin, M.L. (2007). Genomic minimalism in the early diverging intestinal parasite *Giardia lamblia*. *Science*, *317*, 1921-1926.

Olson, P.D.; Zarowiecki, M.; Kiss, F.; Brehm, K. (2012). Cestode genomics - progress and prospects for advancing basic and applied aspects of flatworm biology. *Parasite Immunology*, *34*, 130-150.

Ono, M.; Scott, M.S.; Yamada, K.; Avolio, F.; Barton, G.J.; Lamond, A.I. (2011). Identification of human miRNA precursors that resemble box C/D snoRNAs. *Nucleic Acids Research*, *39*, 3879-3891.

Orom, U.A. & Lund, A.H. (2007). Isolation of microRNA targets using biotinylated synthetic microRNAs. *Methods*, *43*, 162-165.

Poole, C.B.; Davis, P.J.; Jin, J.; McReynolds, L.A. (2010). Cloning and bioinformatic identification of small RNAs in the filarial nematode, *Brugia malayi*. *Molecular and Biochemical Parasitology*, *169*, 87-94.

Prucca, C.G.; Slavin, I.; Quiroga, R.; Elías, E.V.; Rivero, F.D.; Saura, A.; Carranza, P.G.; Luján HD. (2008). Antigenic variation in *Giardia lamblia* is regulated by RNA interference. *Nature*, *456*, 750-754.

Rathjen, T.; Nicol, C.; McConkey, G.; Dalmay, T. (2006). Analysis of short RNAs in the malaria parasite and its red blood cell host. *FEBS Letters*, *580*, 5185-5188.

Rehmsmeier, M.; Steffen, P.; Hochsmann, M.; Giegerich, R. (2004). Fast and effective prediction of microRNA/target duplexes. *RNA*, *10*, 1507-1517.

Robb, S.M.; Ross, E.; Sánchez Alvarado, A. (2008). SmedGD: the *Schmidtea mediterranea* genome database. *Nucleic Acids Research*, *36*, D599-606.

Saraiya, A.A. & Wang, C.C. (2008). snoRNA, a novel precursor of microRNA in *Giardia lamblia*. *PLoS Pathogens*, *4*, e1000224.

Saraiya, A.A.; Li, W.; Wang, C.C. (2011). A microRNA derived from an apparent canonical biogenesis pathway regulates variant surface protein gene expression in *Giardia lamblia*. *RNA*, *17*, 2152-64.

Scott, M.S.; Avolio, F.; Ono, M.; Lamond, A.I.; Barton, G.J. (2009). Human miRNA precursors with box H/ACA snoRNA features. *PLoS Computational Biology*, *5*, e1000507.

Scott, A.L.; Ghedin, E.; Nutman, T.B.; McReynolds, L.A.; Poole, C.B.; Slatko, B.E.; Foster, J.M. (2012). Filarial and Wolbachia genomics. *Parasite Immunology*, *34*, 121-129.

Selbach, M.; Schwanhäusser, B.; Thierfelder, N.; Fang, Z.; Khanin, R.; Rajewsky, N. (2008). Widespread changes in protein synthesis induced by microRNAs. *Nature*, *455*, 58–63.

Saetrom, P. & Snove O. Jr. (2007). Robust machine learning algorithms predict microRNAs genes and targets. *Methods in Enzymology*, *427*, 25-49.

Simões, M.C.; Lee, J.; Djikeng, A.; Cerqueira, G.C.; Zerlotini, A.; da Silva-Pereira, R.A.; Dalby, A.R.; LoVerde, P.; El-Sayed, N.M.; Oliveira, G. (2011). Identification of *Schistosoma mansoni* microRNAs. *BMC Genomics*, *12*, 47

Suh, N. & Blelloch, R. (2011). Small RNAs in early mammalian development: from gametes to gastrulation. *Development, 138*, 1653-1661.

Sun, W.; Julie Li, Y.S.; Huang, H.D.; Shyy, J.Y.; Chien, S. (2010). microRNA: a master regulator of cellular processes for bioengineering systems. *Annual Review of Biomedical Engineering*, *12*, 1-27.

Taft, R.J.; Glazov, E.A.; Lassmann, T.; Hayashizaki, Y.; Carninci, P.; Mattick, J.S. (2009). Small RNAs derived from snoRNAs. *RNA*, *15*, 1233-1240.

Tsai, I.J., Zarowiecki, M., Holroyd, N., Garciarrubio, A., Sanchez-Flores, A., Brooks, K.L., Tracey, A., Bobes, R.J., Fragoso, G., Sciutto, E., Aslett, M., Beasley, H., Bennet, H.M., Cai, J., Camicia, F., Clark, R., Cucher, M., De Silva, N., Day, T.A., Deplazes, P., Estrada, K., Fernández, C., Holland, P.W.H., Hou, J., Hu, S., Huckvale, T., Hung, S.S., Kamenetzky, L., Keane, J.A., Kiss, F., Koziol, U., Lambert, O., Liu, K., Luo, X., Luo, Y., Macchiaroli, N., Nichol, S., Paps, J., Parkinson, J., Pouchkina-Stantcheva, N., Riddiford, N., Rosenzvit, M.C., Salinas, G., Wasmuth, J.D., Zamanian, M., Zheng, Y., The Taenia solium Genome Consortium, Cai, X., Soberón, X., Olson, P.D., Laclette, J.P., Brehm, K., Berriman, M., 2013. The genomes of four tapeworm species reveal adaptations to Parasitism. Nature, in press.

Thomas, S. & Bonchev, D. (2010). A survey of current software for network analysis in molecular biology. *Human Genomics*, *4*, 353-60.

Thomas, M.; Lieberman, J.; Lal, A. (2010). Desperately seeking microRNA targets. *Nature Structural & Molecular Biology*, *17*, 1169-1174.

Vinther, J.; Hedegaard, M.M.; Gardner, P.P.; Andersen, J.S.; Arctander, P. (2006). Identification of miRNA targets with stable isotope labeling by amino acids in cell culture. *Nucleic Acids Research*, *34*, e107.

Wang, Z.; Xue, X.; Sun, J.; Luo, R.; Xu, X.; Jiang, Y.; Zhang, Q.; Pan, W. (2010). An "in-depth" description of the small non-coding RNA population of *Schistosoma japonicum* schistosomulum. *PLoS Neglected Tropical Diseases*, *4*, e596.

Wang, J.; Czech, B.; Crunk, A.; Wallace, A.; Mitreva, M.; Hannon, G.J.; Davis, R.E. (2011). Deep small RNA sequencing from the nematode *Ascaris* reveals conservation, functional diversification, and novel developmental profiles. *Genome Research*, *21*, 1462-1477.

Wang, J.; Liu, X.; Jia, B.; Lu, H.; Peng, S.; Piao, X.; Hou, N.; Cai, P.; Yin, J.; Jiang, N.; Chen, Q. (2012a). A comparative study of small RNAs in *Toxoplasma gondii* of distinct genotypes. *Parasites & Vectors*, *5*, 186.

Wang, C.R.; Xu, M.J.; Fu, J.H.; Nisbet, A.J.; Chang, Q.C.; Zhou, D.H.; Huang, S.Y.; Zou, F.C.; Zhu, X.Q. (2012b). Characterization of microRNAs from *Orientobilharzia turkestanicum*, a neglected blood fluke of human and animal health significance. *PLoS One*, *7*, e47001.

Winter, A.D.; Weir, W.; Hunt, M.; Berriman, M.; Gilleard, J.S.; Devaney, E.; Britton, C. (2012). Diversity in parasitic nematode genomes: the microRNAs of *Brugia pahangi* and *Haemonchus contortus* are largely novel. *BMC Genomics*, *13*, 4.

Xu, M.J.; Liu, Q.; Nisbet, A.J.; Cai, X.Q.; Yan, C.; Lin, R.Q.; Yuan, Z.G.; Song, H.Q.; He, X.H.; Zhu, X.Q. (2010). Identification and characterization of microRNAs in *Clonorchis sinensis* of human health significance. *BMC Genomics*, *11*, 521.

Xu, M.J.; Ai, L.; Fu, J.H.; Nisbet, A.J.; Liu, Q.Y.; Chen, M.X.; Zhou, D.H.; Zhu, X.Q. (2012). Comparative Characterization of MicroRNAs from the Liver Flukes *Fasciola gigantica* and *F. hepatica*. *PLoS One*, *7*, e53387.

Xu, M.J.; Fu, J.H.; Nisbet, A.J.; Huang, S.Y.; Zhou, D.H.; Lin, R.Q.; Song, H.Q.; Zhu, X.Q. (2013). Comparative profiling of microRNAs in male and female adults of *Ascaris suum. Parasitology Research, in press.*

Xue, X.; Zhang, Q.; Huang, Y.; Feng, L.; Pan, W. (2008a). No miRNA were found in *Plasmodium* and the ones identified in erythrocytes could not be correlated with infection. *Malaria Journal, 7*, 47.

Xue, X.; Sun, J.; Zhang, Q.; Wang, Z.; Huang, Y.; Pan, W. (2008b). Identification and characterization of novel microRNAs from *Schistosoma japonicum. PLoS One, 3*, e4034.

Yang, J.S.; Maurin, T.; Robine, N.; Rasmussen, K.D.; Jeffrey, K.L.; Chandwani, R.; Papapetrou, E.P.; Sadelain, M.; O'Carroll, D.; Lai, E.C. (2010). Conserved vertebrate mir-451 provides a platform for Dicer-independent, Ago2-mediated microRNA biogenesis. *Proceedings of the National Academy of Sciences USA, 107*, 15163-15168.

Yang, J.S. & Lai, E.C. (2011). Alternative miRNA biogenesis pathways and the interpretation of core miRNA pathway mutants. *Molecular Cell, 43*, 892-903.

Zerlotini, A.; Aguiar, E.R.; Yu, F.; Xu, H.; Li, Y.; Young, N.D.; Gasser, R.B.; Protasio, A.V.; Berriman, M.; Roos, D.S.; Kissinger, J.C.; Oliveira, G. (2013). SchistoDB: an updated genome resource for the three key schistosomes of humans. *Nucleic Acids Research, 41*, D728-D731.

Zhang, L.; Ding, L.; Cheung, T.H.; Dong, M.Q.; Chen, J.; Sewell, A.K.; Liu, X.; Yates, J.R.3rd.; Han, M. (2007). Systematic identification of *C. elegans* miRISC proteins, miRNAs, and mRNA targets by their interactions with GW182 proteins AIN-1 and AIN-2. *Molecular Cell, 28*, 598-613.

Zhang, Y.Q.; Chen, D.L.; Tian, H.F.; Zhang, B.H.; Wen, J.F. (2009). Genome-wide computational identification of microRNAs and their targets in the deep-branching eukaryote *Giardia lamblia. Computational Biology and Chemistry, 33*, 391-396.

Zheng, Y. (2013). Phylogenetic analysis of the Argonaute protein family in platyhelminths. *Molecular Phylogenetics and Evolution, 66*, 1050-1054.

Zuker, M. (2003). Mfold web server for nucleic acid folding and hybridization prediction. *Nucleic Acids Research, 31*, 3406-3415.

In: MicroRNA and Non-Coding RNA
Editor: James C. Johnson
ISBN: 978-1-62618-442-8

Chapter 4

Functional Roles of Non-Coding RNAs in Glioma and their Clinical Implications

Xiao-Qin Zhang and Gilberto Ka-Kit Leung*
Department of Surgery, Li Ka Shing Faculty of Medicine,
The University of Hong Kong, Hong Kong

Abstract

Glioma is the commonest form of primary brain tumors with poor clinical outcome. MicroRNAs (miRNAs) are endogenous non-coding RNAs (ncRNAs) of approximately 22 nucleotides (nt) that regulate the expression of target messenger RNAs (mRNAs) at the post-transcriptional level. Increasing evidence shows that miRNAs play key roles in the biological behavior of glioma. They are involved in a wide variety of gliomagenic processes including cell proliferation, invasion and apoptosis by acting as tumor suppressors or oncogenes. MiRNAs may potentially be exploited as biomarkers for diagnosis, prognostication, and even therapeutic targets. Recently, the characterization of long non-coding RNAs (lncRNAs) has received considerable attention. Like their smaller non-coding miRNA counterparts, lncRNAs represent a novel class of gene regulators in cancer. Early evidence demonstrates that the differential expression of specific lncRNAs may also serve as novel biomarkers for diagnosis and therapies. In this chapter, we summarize the recent advances in the investigations of miRNAs and lncRNAs in glioma, and discuss their potentials for clinical use.

Keywords: miRNAs, lncRNAs, ncRNAs, glioma, diagnosis, prognosis; therapy

* Corresponding author: Dr. Gilberto K.K. Leung, Division of Neurosurgery, Department of Surgery, Li Ka Shing Faculty of Medicine, The University of Hong Kong, Queen Mary Hospital, 102 Pokfulam Road, Hong Kong. Email: gilberto@hkucc.hku.hk; Phone: +852-2255 3368; Fax: +852-2818 4350.

1. Introduction

Gliomas are the commonest form of primary brain tumor with heterogeneous morphologies and variable prognoses. According to the World Heath Organization (WHO) classification (2007) [1], gliomas include three main histological subtypes based on their cellular origins: astrocytomas (astrocytes), oligodendrogliomas (oligodendrocytes) and oligoastrocytomas (mixture of astrocytes and oligodendrocytes). Within each subtype, gliomas could be further classified into grades I to IV based on the degree of malignancy [1]. Generally, Grade I and II tumors (low-grade gliomas) are relatively benign and carry the optimal clinical outcomes, whereas Grade III and IV tumors (high-grade gliomas) are highly malignant and commonly lethal. The most devastating glioma is grade IV astrocytoma, also known as glioblastoma multiforme (GBM), which is invariably fatal, with a mean patient survival of around 12 to 15 months despite therapeutic interventions with surgery, radiation, and chemotherapy [2]. These tumors are highly invasive and proliferative, and also display marked angiogenesis, necrosis, and resistance to radiation and chemotherapy, which render them one of the most difficult malignancies to treat [2].

Continual discoveries of non-coding RNAs (ncRNAs) have led to profound changes in our understanding of human genetics and cellular regulatory networks [3,4]. NcRNAs are functional RNA molecules that are not translated into proteins. Based on their sizes, ncRNAs are loosely grouped into two major classes: short ncRNAs and long ncRNAs. Short ncRNAs are represented by the well-known microRNAs (miRNAs), a class of single-stranded non-coding RNAs averaged 22 nucleotides (nt) in length [5,6]. MiRNAs have been extensively studied during the past decade and in connection with carcinogenesis [7,8]. It is now recognized that miRNA can regulate target messenger RNA (mRNA) expressions by imperfectly binding to the 3'-UTR sequences of the mRNAs, resulting in the latter's degradation or translational repression [5,6]. In contrast to miRNAs, long ncRNAs (lncRNAs) are mRNA-like transcripts ranging in length from 200nt to ~100kb and lacking significant open reading frames (ORFs). Many identified lncRNAs are transcribed by RNA polymerase II (RNA pol II) and are polyadenylated, but there are exceptions [9,10]. These novel genes are frequently aberrantly expressed in a variety of cancers [11,12], although their functional mechanisms remain largely unknown.

Compelling evidence indicates that miRNAs and lncRNAs play important roles in diverse cellular biological processes and may contribute to cancer development and progression [7,8,11-16]. In glioma, miRNAs have been shown to play key regulatory roles in many biological behaviors [17,18], such as proliferation, invasion, apoptosis, angiogenesis, and stem cell differentiation. Aberrant expressions of specific miRNAs have also been implicated as biomarkers for glioma diagnosis [19-21] and prognostication [21-27]. Similarly, lncRNAs are also emerging as important regulators in the development and progression of a variety of human cancers, although the number of such lncRNAs identified so far is comparatively small [11]. Dysregulation of certain lncRNA may lead to tumor transformation and progression in certain cancer types [11] including glioma [28], and may also be involved in determining tumor metastasis and patient prognosis [29,30]. Some lncRNAs have been found to be detectable in body fluids of cancer patients [31-33], suggesting their potential use as biomarkers.

Table 1. Dysregulated miRNAs in glioma

Authors	Year	Methods	Samples	No. of Up-regulated miRNAs	No. of Down-regulated miRNAs	Reference
Chan et al.	2005	Microarray	GBM vs. NB	5, miR-21 et al.	3	34
Ciafre et al.	2005	Microarray	GBM vs. NB	9, miR-10b, miR-130a, miR-221, miR-25, miR-21 et al.	4, miR-128, miR-181 et al.	35
Silber et al.	2008	RT-PCR	GBM/AA vs. NB	1, miR-10b	12, miR-7, miR-124, miR-124b, miR-137, miR-138, miR-139, miR-218 et al.	36
Godlewski et al.	2008	Microarray	GBM vs. NB	8, miR-21, miR-10b, miR-26a et al.	11, miR-124a, miR-137, miR-139, miR-218, miR-128-2, miR-128-1 et al.	37
Sasayama et al.	2009	Microarray	GBM vs. NB	5, miR-10b, miR-21 et al.	5	38
Dong et al.	2010	Microarray	GBM vs. NB	5, miR-21, miR-25, miR-155, miR-210, miR-130a	12, miR-137, miR-128a, miR-7, miR-124a et al.	39
Conti et al.	2009	RT-PCR	GBM/AA/DA vs. NB	2, miR-21, miR-221	1, miR-181b	40
Wuchty et al.	2011	Microarray	GBM vs. NB	251, miR-21, miR-25, miR-16, miR-10b et al.	24, miR-7, miR-128a, miR-124, miR-181 et al.	41
Li et al.	2011	Microarray	PA/DA/AA vs. NB	3, miR-26a et al.	11, miR-124, miR-128a/b, miR-138, miR-181 et al.	42
Skalsky et al.	2011	Sequencing	GBM vs. NB	19, miR-21, miR-10b, miR-25 et al.	49, miR-124, miR-7, miR-181, miR-138 et al.	43
Malzkorn et al.	2010	RT-PCR	GBM vs. DA	12, miR-16, miR-21, miR-25, miR-210 et al.	2, miR-184 et al.	44
Rao et al.	2010	Microarray	GBM vs. AA	21, miR-21, miR-155, miR-16 et al.	2, miR-128 et al.	45
Guan et al.	2010	RT-PCR	GBM vs. AA	6, miR-21 et al.	10, miR-184, miR-128b et al.	46
Cal et al.	2008	Microarray	CD133+ vs. CD133- GBM cells	6, miR-16 et al.	None found	47
Loftus et al.	2012	Microarray	Migrating vs. non-migrating GBM cells	26	46, miR-16, miR-181 et al.	48

Abbreviations: GBM, glioblastoma multiforme (grade IV); AA, anaplastic astrocytoma (grade III); DA, diffuse astroctyoma (grade II); NB, non-tumoral brain control.

Note: Among the dysregulated miRNAs identified in individual study, only those miRNAs that had been reported in more than one independent study were listed in the table.

In this review, we will focus on the miRNAs and lncRNAs that have been implicated in glioma carcinogenesis and progression. We first provide an overview on profiling studies and functional characterizations of miRNAs in glioma. Thereafter, the known clinical relevance of miRNAs will be summarized, and their potential diagnostic, prognostic and therapeutic uses will be highlighted. Finally, we will review the development as well as the current status of our understanding of the functional roles of lncRNAs in cancers. The specific lncRNAs that have been correlated with glioma biogenesis and progression will be described. In addition, the lncRNA profiling method and the potential utility of lncRNA in cancer clinical practice will also be discussed.

2. miRNAs in Glioma

2.1. Aberrant miRNA Expression Signatures in Glioma

At the time of writing, multiple profiling studies have been reported that broadly characterized the aberrant miRNA expression patterns in glioma. These studies have identified numerous miRNAs that are differentially expressed in gliomas compared to normal tissues, or in gliomas of different malignancy grades or biological behaviors (e.g., migration). Given the large number of dysregulated miRNAs reported and the limit of space, a detailed summary is not feasible, and we have listed here those miRNAs that have been reported in more than one independent study (Table 1).

The first reports of miRNA expression profiling in glioma were independently described by two groups around the same time in 2005 [34,35]. Chan et al. identified five up-regulated and three down-regulated miRNAs out of 180 miRNAs in glioblastoma [34], while Ciafre et al. demonstrated nine up-regulated and four down-regulated miRNAs out of 245 miRNAs [35]. A common miRNA identified by both groups was miRNA-21, which was shown to act as an anti-apoptotic factor [34]. Subsequently, several other studies have also profiled global miRNA expressions in glioma [36-43]. These studies confirmed the over-expression of miR-21 in gliomas, and also identified other commonly dysregulated miRNAs. For example, miR-10b was reported to be up-regulated in glioblastomas compared to normal brain tissues in five independent studies[36-38,41,43]; and miR-25 was up-regulated in three different studies [39,41,43]. Some of these studies also analyzed the regulatory interactions between miRNAs and mRNAs [39,41]. Taken together, over 10 miRNA profiling studies have been reported in gliomas to date. It is interesting to note that the findings from these reports showed marked disparity, the degree of concordance between studies being very low. For example, among the 440 miRNAs identified to be dysregulated in gliomas compared to non-tumoral brain tissues, fewer than 20 miRNAs appeared in more than two independent studies (Table 1). The composition of the miRNA arrays used, the diversities of analytic methodology and control samples may underlie this disparity.

Other studies explored the differential expressions of miRNA in gliomas of different malignancy grades. For example, Malzkorn et al. analyzed the miRNA expression patterns in glioma patients who had the progression from WHO grade II to IV tumors [44]. This revealed the differential expression of 14 miRNAs, including 12 miRNAs that were up-regulated and two that were down-regulated during progression (Table 1). Another study compared miRNA expressions between grade III and grade IV astrocytomas (GBM), and revealed the differential expression of a set of 23 miRNAs [45]. Moreover, by using this 23-miRNA signature, the grade III tumors could be discriminated from grade IV tumors with an overall accuracy of 89.7% [45]. Similarly, Guan et al. examined the global miRNA expression pattern between grade III astrocytomas and GBM, and suggested that 16 miRNAs were associated with malignant progression [46]. These findings have contributed to a better understanding of the molecular mechanisms involved in glioma tumorigenesis and malignant progression.

A unique study attempted to assess differential expression levels of miRNAs between GBM stem cells (CD133+) and non-stem cells (CD133-) [47]. Interestingly, several miRNAs were found to be over-expressed in the non-stem cell population, including miRNA-451, miR-486, miR-16, miR-107, and miR-185, but no others were found to be down-regulated in this cell population. The authors hypothesized that the over-expression of these miRNAs in the non-stem cell population had driven stem cells to differentiate and lose their stem cell characteristics, and increasing the levels of these miRNAs in the stem cell population may potentially inhibit cell growth and tumor proliferation by reducing the self-renewal potential. In a more recent study, the global expression of miRNAs between migrating and non-migrating glioblastoma cells was analyzed [48]. It was found that 72 miRNAs (26 up-regulated and 46 down-regulated) were significantly dysregulated in the migrating and migration-restricted tumor cells [48].

2.2. Functional Roles of miRNAs in Glioma

Discoveries of aberrant miRNAs in gliomas have prompted the investigations into their functions. It is now evident that miRNAs can directly target and regulate the key proteins that function at different steps of the core signaling pathways of glioma, thereby controlling a wide range of biological functions such as proliferation, apoptosis, invasion, angiogenesis and stem cell behavior [17,49]. Depending on their target genes, miRNAs can be oncogenic or tumor-suppressive. Mounting evidence has shown that the dysregulation of miRNAs may potentially affect their mRNA targets hence the corresponding cellular signaling pathways, leading to effects on glioma biological behavior. Given that the detailed description of miRNAs functions in glioma signaling pathways have already been comprehensively discussed in several recently published reviews [17,18,50,51], we provide here only a brief summary of the key miRNAs that have been implicated in regulating glioma proliferation, apoptosis, migration, invasion, angiogenesis and also stem cell behavior (Table 2). However, it should be emphasized that many of these phenomena do not exist in isolation but exhibit complex interplays that ultimately account for tumorigenetic events and progression in glioma.

2.3. miRNAs and Glioma Treatment Resistance

Comparisons between treatment-responsive and treatment-resistant glioma cells have revealed significant differences in miRNA expression patterns between the two groups [52], suggesting that miRNAs may have roles in the development of treatment resistance. And indeed, differential miRNA expression could provide not only valuable information about sensitivity or resistance to treatment prior to the administration of therapy but also potential therapeutic targets for the enhancement of treatment response.

Table 2. Functional roles of miRNAs in glioma

miRNA	Tumor/ Normal	Reported targets	Function in glioma
miR-10b	Up	HOX10D, Bim, TFAP2C, p16,p21	↑ I, M; ↓ A
miR-21	Up	PDCD4, PTEN, RECK, TPM1, TIMP3	↑ I, P, M; ↓ A
miR-25	Up	Bim, p53, p57, KLF4,	↑ I, P, M; ↓ A
miR-26a	Up	PTEN, CDK6, Cyclin E1, EZH2, CPEB2/3/4, BDNF	↑ I, P, M; ↓ A
miR-125b	Up	BMF, CDK6, CDC25A, p53	↑ P; ↓ A
miR-221/222	Up	p27, NIAP, BIRC1	↑ P, I
miR-7	Down	EGFR, IRS-1, IRS-2	↑ A; ↓ I, P, M, GSCs
miR-15b	Down	Cyclin E1, NRP-2	↑ A; ↓ I, AG
miR-34a	Down	c-Met,Notch1, Notch2 and CDK6	↑ A; ↓ P, M, GSCs
miR-128	Down	EGF receptor, Bmil, E2F3a, ARP5, PDGFRa, WEE1,	↑ A; ↓ P, M, GSCs
miR-137	Down	CDK6; Cox2	↓ P, I, GSCs
miR-181a/b/c	Down	Bcl-2	↑ A; ↓ I
miR-218	Down	IKK-b, MMP-9	↓ M, I

Abbreviations: I, invasion; A, apoptosis; P, proliferation; M, migration; GSCs, glioma stem cells; AG, angiogenesis.

In glioma, resistance of tumor cells to radiotherapy plus concomitant and adjuvant temozolomide (TMZ) chemotherapy continues to be a major clinical obstacle to the successful treatment of this disease. In view of the fact that TMZ is now a standard agent for malignant glioma, several studies have attempted to identify miRNAs specifically involved in TMZ resistance. In one study, the authors first established resistant U251 variants from the parental U251 glioma cells, and then performed miRNA microarray analysis on both cell populations [52]. The results showed that three miRNAs (miR-195, miR-455-3p, and miR-10a*) were significantly up-regulated in the resistant cells. Suppression of miR-455-3p or miR-10a* in combination with TMZ showed modest cell killing effect, while knock down of miR-195 in the presence of TMZ strongly enhanced the effect. Another study examined the correlation between miRNA expression levels and patients' responses to concomitant TMZ therapy [53]. They found that miR-181b and miR-181c were significantly down-regulated in patients who responded to TMZ in comparison to patients with progredient disease, suggesting that the expression levels of miR-181b and miR-181c may serve as a predictive marker of response to TMZ [53]. In our previous study, we investigated miR-21 expression level and the effect of miR-21 inhibition in D54 glioblastoma cells with acquired TMZ resistance [54]. We found that chronic TMZ exposure led to acquired TMZ resistance and elevated miR-21 expression. Concomitant use of miR-21 inhibition and TMZ resulted in significantly apoptotic effects than TMZ treatment alone [54]. The effect of miR-21 expression on TMZ response was also observed by several other studies [55,56]. This miRNA could contribute to tumor resistance by decreasing Bax/Bcl-2 ratio and caspase-3 activity [56]. These findings suggest that miRNAs play a role in chemotherapy resistance and could be a novel target for GBM treatment. The miRNAs that have been implicated in glioma chemotherapy resistance were summarized in Table 3.

Table 3. miRNAs associated with treatment resistance in glioma

miRNA	Cell line	Effects on treatment	Comments	Reference
miR-10a*	U251	Chemo	Down-regulation increased sensitivity to TMZ	52
miR-195	U251	Chemo	Down-regulation increased sensitivity to TMZ	52
miR-455-3p	U251	Chemo	Down-regulation increased sensitivity to TMZ	52
miR-181b/c	tissues	Chemo	Down-regulated in TMZ-responsive GBM patients	53
miR-21	D54	Chemo	Down-regulation increased sensitivity to TMZ	54
miR-21	U251	Chemo	Down-regulation increased sensitivity to TMZ	55
miR-21	U87	Chemo	Down-regulation increased sensitivity to TMZ	56
miR-221/ 222	U251	Radio	Down-regulation increased radio-sensitivity	57
miR-181a	U87	Radio	Over-expression increased radio-sensitivity	58
miR-7	U251/U87 /GSCs	Radio	Over-expression increased radio-sensitivity	59
miR-24-1* et al.	LN229/T98G/U87	Radio	Dysregulated after radiation treatment	60
miR-125b-2	GSCs	Chemo	Down-regulation increased sensitivity to TMZ	64

Abbreviations: GSCs, glioma stem cells.

Other miRNAs were linked with resistance to radiotherapy in glioma. Zhang et al. found that inhibition of miR-221/222 enhanced the radio-sensitivity of U251 glioma cells through the up-regulation of p27 (kip1) [57]. Chen et al. observed that over-expression of miR-181a significantly sensitized U87 glioma cells to radiation by targeting Bcl-2 [58]. Ectopic over-expression of miR-7 was found to increase radio-sensitivity in U251 and U87 malignant glioma cells by attenuating EGFR and AKT expression [59]. In a more recent study, the miRNA expression profiles were analyzed in three glioma cell lines (LN229, T98G, U87) before and after radiation treatment, respectively [60]. The results showed that ionizing radiation induced significant changes in miRNA expression profiles. Further bioinformatic analysis revealed that the majority of these dysregulated miRNAs were involved in the regulation of key cellular processes like apoptosis, proliferation and invasion, which partially explained their roles in radio-resistance. The effects of miRNAs modulation on radio-sensitivity of glioma cell lines are summarized in Table 3.

Since glioma cancer stem cells (GSCs) appear to be the major contributors to tumor initiation and chemo/radio resistance in glioma, identifying miRNAs that are involved in regulating GSCs self-renewal and differentiation may be helpful for the development of novel treatment approaches. Kefas et al. reported that miR-7 was involved in regulating the malignant parameters in GSCs [61]. This was the first report that implicated a miRNA in the regulation of GSCs. Since then, two studies uncovered the critical roles for miRNA-34a in GSCs self-renewal [62,63], and that over-expression of miR-34a induced GSCs differentiation. In two more recent studies, Shi et al. [64] found that miR-125b-2 was over-expressed in GBM tissues and the corresponding stem cells; down-regulation of miR-125b-2 expression in GSCs could allow TMZ to induce GSCs apoptosis [64]. Chan et al. reported the vital role of miR-138 in promoting growth and survival of GSCs [65]. These studies provide a better understanding of the mechanisms of glioma initiation and progression, and perhaps also for the use of miRNAs as future targets for effective glioma therapy.

2.4. miRNAs as Biomarkers for Glioma Diagnosis and Prognosis

Following the identification of miRNA expression differences in gliomas compared with normal brain tissues, or in gliomas with different malignancy grades, the investigation of their roles as biomarkers for glioma diagnosis and grading is a logical next step. Moreover, differential miRNA expressions in patients with different clinical outcomes indicated that they may also act as prognosis markers.

Expression levels of miR-182 [23], miR-196 [46] and miR-17 [24] were found to be significantly associated with advanced pathological grades and poor prognosis in glioma. In another study, low levels of miR-181b and miR-106a, or high expression of miR-21, were significantly associated with poor patient survival [25]. Analysis of the miRNA expression data in GBM patients (n=222) derived from the publicly available The Cancer Genome Atlas (TCGA) identified a ten-miRNA signature that could predict GBM patient survival [22]. Moreover, multivariate analysis showed that the prognostic value of this signature was independent of patients' age [22]. Niyazi et al. reported the prognostic value of a set of 30 miRNAs that could separate GBM patients into two subgroups with distinct prognoses [26]. Another study identified miR-181d as a prognostic factor in TMZ-treated GBM patients [27]. More interestingly, they found that miR-181d played its predictive role for TMZ response by targeting methyl-guanine-methyl-transferase (MGMT), an enzyme that can significantly affect TMZ cytotoxicity [27]. This was the first miRNA reported to target MGMT expression directly in glioma.

The discovery of miRNAs in the circulating system of glioma patients opened up a new and exciting possibility of using miRNAs as readily available biomarkers for diagnosis and prognostications in glioma (Table 4). For example, Teplyuk et al. demonstrated that miRNAs could be reliably detected in the cerebrospinal fluid (CSF) of brain tumor patients, and the use of as few as 7 miRNAs could discriminate GBM from metastatic brain cancers with high accuracy (91%-99%) [20]. Roth et al. showed that a specific miRNA signature in the peripheral blood of GBM patients allowed for the discrimination between GBM patients and healthy controls with an accuracy of 81% [19]. Wang et al. found that the plasma levels of three miRNAs (miR-21, miR-128 and miR-342-3p) were significantly altered in GBM patients compared to normal controls, and the differential expression of these three miRNAs could discriminate glioma from healthy controls with high specificity and sensitivity [21].

These findings suggest that miRNAs have potential uses as novel biomarkers in the clinical management of glioma patients.

2.5. miRNA as Therapeutic Targets in Glioma

The fact that miRNA dysregulation in glioma has a pathogenic effect provides the rationale for using miRNA as potential therapeutic targets in glioma. Typically, the use of miRNAs in glioma therapy (and also other cancers) involves two approaches. The first approach is to inhibit oncogenic miRNAs with a gain of function using miRNA antagonists. The second approach is to restore tumor suppressive miRNAs with a loss of function using miRNA mimics.

Several reports have shown the potential use of miRNA-based approach in glioma therapy. MiR-21, being one of the most up-regulated miRNAs in gliomas, with anti-apoptotic and pro-invasive functions, has been selected as one of the first targets for therapeutic interventions. Cortsten et al. studied the combined effect of suppressing miR-21 and neural precursor cells (NPCs)-mediated secretable variant of the cytotoxic agent tumor necrosis factor-related apoptosis inducing ligand (S-TRAIL) in gliomas (U87) [66]. They found that miR-21 knock down disrupted glioma growth *in vitro* and displayed synergistic cytotoxicity with NPCs delivered S-TRAIL in glioma cell lines in human gliomas [66]. This was confirmed also *in vivo*. Consistent with Corsten's results, Zhou et al. demonstrated that injecting miR-21 antisense inhibitors into mice bearing subcutaneous U251 tumors resulted in significant decrease in tumor volumes [67]. In another study involving a tumor suppressor miRNA, miR-34a, re-introducing the latter into U87 glioma cells led to strong inhibition of tumor growth in xenograft models [62]. These results indicated the potential roles of miRNAs as targets for therapeutic intervention in glioma.

Although miRNA-based therapies have been shown to effectively inhibit tumor growth *in vitro* and *in vivo*, many challenges remain before these can be translated into clinical practice. The first challenge is the effective delivery of miRNA antagonists or mimics. Although the transient delivery of miRNA antagonists or mimics using lipid emulsion exhibited tumor-inhibitory effects on some glioma cell lines and mice xenograft models, the transfection efficiency could be extremely low for some glioma cell lines. At this point in time, although the virus-based delivery approach showed high gene transfer efficiency and prolonged therapeutic time, information on its bio-safety is lacking. Therefore, other delivery techniques, and pharmaceutical formulations that can increase the efficiency and safety of gene delivery are still urgently needed. The second challenge is to maintain target specificity. This is especially true for miRNA-based therapies since a single miRNA may regulate hundreds of target mRNA genes and off-target gene silencing commonly occur in miRNA regulation [68], which only requires partial complementary binding between miRNA and mRNA. Therefore, it is important to evaluate the effect of a specific miRNA-mediated therapy on a large proteome-wide scale to prevent unwanted gene alteration, and treatment-related side-effects.

3. lncRNAs in Glioma

3.1. Discovery of lncRNAs

The earliest discovery of lncRNAs predated the discovery of miRNAs, but they have been considered to be "transcriptional noise" or "Junk RNA" of the human genome and did not receive much attention. One of the first lncRNA genes reported was the imprinted H19 gene [69]. This was soon followed by the discovery of another lncRNA, XIST [70]. The term "lncRNA" had not been coined at the time, and the discovery of the first miRNA, lin-14 [71], dramatically redirected the research focus from lncRNAs to miRNAs. The variety and dynamics of lncRNA expression was not fully appreciated until the introduction of whole transcriptome sequencing. With the advancement of large-scale complementary DNA (cDNA) sequencing projects such as FANTOM (Functional Annotation of Mammalian cDNA), it was revealed that the mammalian genome is extensively transcribed, although a large portion of this represented non-coding sequences [72,73]. Coupled with the novel functional annotations of a handful lncRNAs, this discovery promoted researches focusing on lncRNA discovery and characterization. Currently, it is estimated that the lncRNA genes in the human genome range from 7,000 to 23,000, implying that this class of ncRNA will represent an enormous, yet undiscovered, component of normal cellular networks that may be disrupted in cancer biology [74].

Table 4. miRNAs detected in the circulating system of glioma patients

Authors	Circulating medium	miRNAs detected	Clinical-related phenotypes	Reference
Teplyuk et al.	CSF	miR-21, miR-10b, miR-125b, miR- 200 a/b/c, miR-141	Differential expression discriminates glioma from metastatic brain tumor.	20
Roth et al.	Peripheral blood	miR-128, miR-342-3p et al.	Differential expression discriminates glioma from healthy controls.	19
Wang et al.	plasma	miR-21, miR-128, miR-342-3p	Differential expression discriminates glioma from healthy controls.	21

Abbreviations: CSF, cerebrospinal fluid (CSF).

3.2. Emerging role of lncRNAs in Cancinogenesis

A role for differential lncRNA expression in cancer had been suspected for many years but strong supporting evidence was lacking. With advancements in cancer transcriptome profiling and accumulating evidence supporting lncRNA function, a number of aberrantly expressed lncRNAs were found to be associated with cancer disease process [11]. lncRNAs have been implicated to regulate a range of biological functions and the disruption of some of these functions may lead to cancer development and progression. The aberrant expression of specific lncRNAs has also been linked with gliomagenesis [75,76]. The lncRNAs that have been implicated in cancer, and also other diseases, could be found in the public database lncRNAdb (a database of literature described lncRNAs) [77].

3.3. lncRNA Profiling Method

Like studies on other gene classes in cancer biology, lncRNA expression differences between tumor and normal samples were among the first types of evidence that we seek. Although some of the specific lncRNAs whose aberrant expressions and functions have been identified, the global lncRNA expression profile in most cancers, including glioma, remains largely unknown.

To investigate their global expression profile, a comprehensive profiling method is essential. Generally, lncRNA profiling could be achieved through two technologies: array-based profiling and sequencing-based profiling. Both methods are being employed in lncRNA profile study. For example, by using a custom-designed microarray platform, Mercer et al. examined the dynamic lncRNA expression accompanying neuronal and glial cell differentiation [78]. By using a commercial microarray special for lncRNA, Yang et al. analyzed the differential lncRNA expression between HBV-related hepatocellular carcinoma (HCC) and paired peritumoral tissues [79]. On the other hand, Lin et al. used next generation sequencing (RNA-Seq) to examine the dynamic changes in the expression of lncRNAs and protein-coding genes during the transition from induced pluripotent stem cells (iPSCs) to early differentiated neurons [80]. They found that a number of lncRNA genes underwent radical changes in expression during this transition [80].

These studies showed the great potential of applying commercial profiling technology as a powerful tool to study lncRNA profile. However, from most researchers' points of view, one of the main hurdles is the cost. This is especially true for the clinical sample-based profiling studies, which commonly involved tens of or even hundreds of samples to be tested because of the high variance between specimens. To address this, some investigators have begun to mine previously published microarray gene expression data or sequencing data for lncRNA profiles. Previous studies have generated abundant high-throughput data which are accessible in public data bases such as the Gene Expression Omnibus (GEO). lncRNA profiling could be achieved by re-analyzing these data since a large number of lncRNAs were fortuitously detected and preserved in previous studies. This method is feasible and attractive, not only in its accuracy, low cost and ready availability, but also in that it allows the analysis of mRNA and lncRNA expressions at the same time. By using the previously published serial analysis of gene expression (SAGE) libraries deposited in GEO, for example, Gibb et al. analyzed the lncRNA expression profiles across a broad spectrum of normal human tissues and cancers [81]. In our recent study, we applied a lncRNA classification pipeline to mine lncRNA expression profiles from previously published glioma gene expression data profiled through Affymetrix HG-U133 plus 2.0 microarray platform [76]. Two other similar studies mined lncRNA expression profile from Affymetrix HG-U133A and B arrays [82,83]. All these studies highlight the feasibility and effectiveness of using the mining approach in lncRNA expression profile studies.

3.4. Aberrant lncRNA Expression Signatures in Glioma

By using the lncRNA-minging approach mentioned above, we have analyzed the aberrant lncRNA expression patterns in two large cohorts of glioma data sets downloaded from the GEO [76]. We compared the lncRNA gene expression profiles in tumors relative to normal brain controls, as well as in tumors with different malignancy grades or histological cell subtypes. We observed that lncRNAs were aberrantly expressed in gliomas as compared to normal brain tissues. Moreover, specific lncRNAs dysregulation was found to be closely associated with malignancy progression in glioma, since their expressions showed step-wise increase or decrease with ascending malignancy scale. We also found that lncRNAs were differentially expressed between tumors with different histological subtypes (astrocytic versus oligodendroglial tumors), and the differential expression of a 23-lncRNA signature may help to distinguish these two types of tumor with high accuracy rate. Of the sets of dysregulated lncRNAs we identified, CRNDE is a particularly interesting candidate. It was found to be the most highly up-regulated lncRNA in gliomas compared to normal brain tissues, and correlated positively with tumor grading. The significant up-regulation of CRNDE was also observed in a variety of other cancers [84], suggesting that it may be involved in the core cellular signaling pathways that are common to diverse types of cancers.

Aberrant lncRNA expressions in glioma were also observed by another group. Han et al. analyzed the lncRNA and mRNA profiles in GBM and normal brain tissues by using a commercial lncRNA array [75]. They found that lncRNAs expression profile in GBM tissues was significantly altered compared to normal brain tissues. Further lncRNA gene network analysis revealed that two aberrantly expressed lncRNAs (ASLNC22381 and ASLNC2081) were likely to play roles in the regulation of glioma signaling pathways. These two lncRNAs may target insulin-like growth factor-1(IGF-1) in the recurrence and malignant progression of GBM.

3.5. Functional Roles of lncRNA in Gliomagenesis

Only a limited number of published studies have addressed directly lncRNAs' regulation and function in glioma. However, these studies represent the beginnings of a new direction of lncRNA-based research that will unravel the molecular complexity underlying glioma pathogenesis. The maternally expressed gene 3 (MEG3) is one of the first lncRNAs proposed to play important functional roles in glioma. MEG3 was implicated as a tumor suppressor lncRNA due to its ability to stimulate both p53-dependent and p53-independent apoptotic pathways. MEG3 is expressed in many normal human tissues, with the highest expression in the brain and pituitary gland, but is not detectable in various cancer cell lines [85,86]. This suggests a potential role of this lncRNA in suppressing cell growth. In gliomas, it was demonstrated that MEG3 expression was markedly decreased in tumor tissues compared with adjacent normal tissues [28]. Moreover, ectopic expression of MEG3 inhibited cell proliferation and promoted cell apoptosis in U251 and U87 human glioma cell lines. Further evidence revealed that MEG3 was associated with the tumor-suppressive gene p53, and this association was required for p53 activation [28]. These data suggest an important role of MEG3 in the molecular etiology of glioma.

3.6. Perspectives of lncRNAs in Cancer Clinical Practice

The marked increase or decrease in lncRNA expression levels in tumors compared with normal tissues, or in tumors with different clinical features, suggested that lncRNAs may also be exploited as biomarkers for diagnosis, prognostication, or even therapeutic agents. Indeed, the prognostic values of specific lncRNAs have been reported in several other cancers. One of the first lncRNAs described to have prognostic significance in cancer was MALAT1 [87]. This lncRNA was found to be highly up-regulated in non-small cell lung cancer patients, and was associated with high metastatic potential and poor patient prognosis [87].

Another well-recognized prognostic lncRNA was HOTAIR, which was highly up-regulated in breast tumors, and correlated with both metastatic potential and poor survival [29]. lncRNA HEIH was reported to be associated with recurrence and prognosis in hepatocellular carcinoma (HCC) patients [79]. In glioma, the intriguing relationship between lncRNA profile and GBM patients' prognosis is currently being investigated in our laboratory.

Like miRNA, lncRNAs could be detected in the body fluids of cancer patients by using simple polymerase chain reaction (PCR) method, suggesting that lncRNA may also be exploited as non-invasive biomarkers in clinical practice. For example, the prostate specific lncRNA PCA3 (or DD3), has already been developed into a highly specific biomarker in clinical trials [32,33]. PCA3 could be consistently detected in the urine of prostate cancer patients, and demonstrated higher diagnostic specificity compared with the standard prostate marker prostate specific antigen (PSA) [32,33]. Similarly, lncRNA CRNDE could be detected in the plasma of colorectal cancer patients. The large difference in its expression showed promise in discriminating between colorectal adenomas and cancers from healthy controls [31].

lncRNAs might also be useful as therapeutic agents. Like mRNA or miRNA genes, lncRNAs expression level could be manipulated by the commonly used transgene approaches such as the RNA interference (RNAi) or gene over-expression vector, indicating they may also be used as potential targets for therapeutic intervention.

To date, very few studies involving lncRNAs as therapeutic agents in cancers have been performed. In breast cancer, knock-down of lncRNA HOTAIR using RNAi technology has been shown to effectively limit the metastatic potential of breast cancer cells in mice xenograft model, while enforcing expression of HOTAIR using expression vector promoted cancer invasiveness [29].

A plasmid construct, harboring a diphtheria toxin gene driven by lncRNA H19 gene promoter, has been investigated for the treatment of a series of cancers and yielded encouraging results [88]. These evidence indicate the potential role of lncRNAs as cancer therapy targets.

Conclusion

Advances in non-coding RNA (ncRNA) research have provided compelling evidence that miRNAs and lncRNAs play key roles in glioma biogenesis and progression. However, for the majority of ncRNA genes, especially lncRNA genes, their functional roles are still incompletely understood.

Differential expression of specific genes is becoming recognized as promising biomarkers that may help distinguish patients with different clinical features. Some of these genes represent attractive drug targets since they may regulate expression of key proteins in glioma cells. However, many issues need to be addressed prior to clinical translation including the delivery of gene interference agents, target specificity and biosafety.

References

[1] Louis, DN; Ohgaki, H; Wiestler, OD; Cavenee, WK; Burger, PC; et al. (2007). The 2007 WHO classification of tumours of the central nervous system. *Acta Neuropathol*, 114, 97-109.

[2] Clarke, J; Butowski, N; Chang, S. (2010). Recent advances in therapy for glioblastoma. *Arch Neurol*, 67, 279-283.

[3] Diederichs, S. (2012). Non-coding RNA and disease. *RNA Biol*, 9, 701-702.

[4] Mattick, JS; Makunin, IV. (2006). Non-coding RNA. *Hum Mol Genet*, 15 Spec No 1, R17-29.

[5] Ambros, V. (2004). The functions of animal microRNAs. *Nature*, 431, 350-355.

[6] Filipowicz, W; Bhattacharyya, SN; Sonenberg, N (2008). Mechanisms of post-transcriptional regulation by microRNAs: are the answers in sight? *Nat Rev Genet*, 9, 102-114.

[7] Farazi, TA; Spitzer, JI; Morozov, P; Tuschl, T. (2011). miRNAs in human cancer. *J Pathol*, 223, 102-115.

[8] Sassen, S; Miska, EA; Caldas, C. (2008). MicroRNA: implications for cancer. *Virchows Arch*, 452, 1-10.

[9] Cheng, J; Kapranov, P; Drenkow, J; Dike, S; Brubaker, S; et al. (2005). Transcriptional maps of 10 human chromosomes at 5-nucleotide resolution. *Science*, 308, 1149-1154.

[10] Wu, Q; Kim, YC; Lu, J; Xuan, Z; Chen, J; et al. (2008). Poly A- transcripts expressed in HeLa cells. *PLoS One*, 3, e2803.

[11] Gibb, EA; Brown, CJ; Lam, WL. (2011). The functional role of long non-coding RNA in human carcinomas. *Mol Cancer*, 10, 38.

[12] Gutschner, T; Diederichs, S. (2012). The hallmarks of cancer: a long non-coding RNA point of view. *RNA Biol*, 9, 703-719.

[13] Galasso, M; Sana, ME; Volinia, S. (2010). Non-coding RNAs: a key to future personalized molecular therapy? *Genome Med*, 2, 12.

[14] Lee, SK; Calin, GA. (2011). Non-coding RNAs and cancer: new paradigms in oncology. *Discov Med*, 11, 245-254.

[15] Yogev, O; Lagos, D. (2011). Noncoding RNAs and cancer. Silence 2, 6.

[16] Yang, W; Lee, DY; Ben-David, Y. (2011). The roles of microRNAs in tumorigenesis and angiogenesis. *Int J Physiol Pathophysiol Pharmacol*, 3, 140-155.

[17] Silber, J; James, CD; Hodgson, JG. (2009). microRNAs in gliomas: small regulators of a big problem. Neuromolecular Med 11, 208-222.

[18] Chiocca, EA; Lawler, SE. (2010). The many functions of microRNAs in glioblastoma. *World Neurosurg*, 73, 598-601.

[19] Roth, P; Wischhusen, J; Happold, C; Chandran, PA; Hofer, S; et al. (2011). A specific miRNA signature in the peripheral blood of glioblastoma patients. *J Neurochem*, 118, 449-457.

[20] Teplyuk, NM; Mollenhauer, B; Gabriely, G; Giese, A; Kim, E; et al. (2012). MicroRNAs in cerebrospinal fluid identify glioblastoma and metastatic brain cancers and reflect disease activity. *Neuro Oncol*, 14, 689-700.

[21] Wang, Q; Li, P; Li, A; Jiang, W; Wang H; et al. (2012). Plasma specific miRNAs as predictive biomarkers for diagnosis and prognosis of glioma. *J Exp Clin Cancer Res*, 31, 97.

[22] Srinivasan, S; Patric, IR; Somasundaram, K. (2011). A ten-microRNA expression signature predicts survival in glioblastoma. *PLoS One*, 6, e17438.

[23] Jiang, L; Mao, P; Song, L; Wu, J; Huang, J; et al. (2010). miR-182 as a prognostic marker for glioma progression and patient survival. *Am J Pathol*, 177, 29-38.

[24] Lu, S; Wang, S; Geng, S; Ma, S; Liang, Z; et al. (2012). Increased Expression of microRNA-17 Predicts Poor Prognosis in Human Glioma. *J Biomed Biotechnol*, 2012, 970761.

[25] Zhi, F; Chen, X; Wang, S; Xia, X; Shi, Y; et al. (2010). The use of hsa-miR-21, hsa-miR-181b and hsa-miR-106a as prognostic indicators of astrocytoma. *Eur J Cancer*, 46, 1640-1649.

[26] Niyazi, M; Zehentmayr, F; Niemoller, OM; Eigenbrod, S; Kretzschmar, H; et al. (2011). MiRNA expression patterns predict survival in glioblastoma. *Radiat Oncol*, 6, 153.

[27] Zhang, W; Zhang, J; Hoadley, K; Kushwaha, D; Ramakrishnan, V; et al. (2012). miR-181d: a predictive glioblastoma biomarker that downregulates MGMT expression. *Neuro Oncol*, 14, 712-719.

[28] Wang, P; Ren, Z; Sun, P. (2012). Overexpression of the long non-coding RNA MEG3 impairs in vitro glioma cell proliferation. *J Cell Biochem*, 113, 1868-1874.

[29] Gupta, RA; Shah, N; Wang, KC; Kim, J; Horlings, HM; et al. (2010). Long non-coding RNA HOTAIR reprograms chromatin state to promote cancer metastasis. *Nature*, 464, 1071-1076.

[30] Tahira, AC; Kubrusly, MS; Faria, MF; Dazzani, B; Fonseca, RS; et al. (2011). Long noncoding intronic RNAs are differentially expressed in primary and metastatic pancreatic cancer. *Mol Cancer*, 10, 141.

[31] Graham, LD; Pedersen, SK; Brown, GS; Ho, T; Kassir, Z; et al. (2011). Colorectal Neoplasia Differentially Expressed (CRNDE), a Novel Gene with Elevated Expression in Colorectal Adenomas and Adenocarcinomas. *Genes Cancer*, 2, 829-840.

[32] Tinzl, M; Marberger, M; Horvath, S; Chypre, C. (2004). DD3PCA3 RNA analysis in urine--a new perspective for detecting prostate cancer. *Eur Urol*, 46, 182-186; discussion 187.

[33] Hessels, D; Klein Gunnewiek, JM; van Oort, I; Karthaus, HF; van Leenders, GJ; et al. (2003). DD3(PCA3).-based molecular urine analysis for the diagnosis of prostate cancer. *Eur Urol*, 44, 8-15; discussion 15-16.
[34] Chan, JA; Krichevsky, AM; Kosik, KS. (2005). MicroRNA-21 is an antiapoptotic factor in human glioblastoma cells. *Cancer Res*, 65, 6029-6033.
[35] Ciafre, SA; Galardi, S; Mangiola, A; Ferracin, M; Liu, CG; et al. (2005). Extensive modulation of a set of microRNAs in primary glioblastoma. *Biochem Biophys Res Commun*, 334, 1351-1358.
[36] Silber, J; Lim, DA; Petritsch, C; Persson, AI; Maunakea, AK; et al. (2008). miR-124 and miR-137 inhibit proliferation of glioblastoma multiforme cells and induce differentiation of brain tumor stem cells. *BMC Med*, 6, 14.
[37] Godlewski, J; Nowicki, MO; Bronisz, A; Williams, S; Otsuki, A; et al. (2008). Targeting of the Bmi-1 oncogene/stem cell renewal factor by microRNA-128 inhibits glioma proliferation and self-renewal. *Cancer Res*, 68, 9125-9130.
[38] Sasayama, T; Nishihara, M; Kondoh, T; Hosoda, K; Kohmura, E. (2009). MicroRNA-10b is overexpressed in malignant glioma and associated with tumor invasive factors, uPAR and RhoC. *Int J Cancer*, 125, 1407-1413.
[39] Dong, H; Siu, H; Luo, L; Fang, X; Jin, L; et al. (2010). Investigation gene and microRNA expression in glioblastoma. *BMC Genomics*, 11 Suppl, 3, S16.
[40] Conti, A; Aguennouz, M; La Torre, D; Tomasello, C; Cardali, S; et al. (2009). miR-21 and 221 upregulation and miR-181b downregulation in human grade II-IV astrocytic tumors. *J Neurooncol*, 93, 325-332.
[41] Wuchty, S; Arjona, D; Li, A; Kotliarov, Y; Walling, J; et al. (2011). Prediction of Associations between microRNAs and Gene Expression in Glioma Biology. *PLoS One*, 6, e14681.
[42] Li, D; Chen, P; Li, XY; Zhang, LY; Xiong, W; et al. (2011). Grade-specific expression profiles of miRNAs/mRNAs and docking study in human grade I-III astrocytomas. OMICS 15, 673-682.
[43] Skalsky, RL; Cullen, BR. (2011). Reduced expression of brain-enriched microRNAs in glioblastomas permits targeted regulation of a cell death gene. *PLoS One*, 6, e24248.
[44] Malzkorn, B; Wolter, M; Liesenberg, F; Grzendowski, M; Stuhler, K; et al. (2010). Identification and functional characterization of microRNAs involved in the malignant progression of gliomas. *Brain Pathol*, 20, 539-550.
[45] Rao, SA; Santosh, V; Somasundaram, K. (2010). Genome-wide expression profiling identifies deregulated miRNAs in malignant astrocytoma. *Mod Pathol*, 23, 1404-1417.
[46] Guan, Y; Mizoguchi, M; Yoshimoto, K; Hata, N; Shono, T; et al. (2010). MiRNA-196 is upregulated in glioblastoma but not in anaplastic astrocytoma and has prognostic significance. *Clin Cancer Res*, 16, 4289-4297.
[47] Gal, H; Pandi, G; Kanner, AA; Ram, Z; Lithwick-Yanai, G; et al. (2008). MIR-451 and Imatinib mesylate inhibit tumor growth of Glioblastoma stem cells. *Biochem Biophys Res Commun*, 376, 86-90.
[48] Loftus, JC; Ross, JT; Paquette, KM; Paulino, VM; Nasser, S; et al. (2012). miRNA expression profiling in migrating glioblastoma cells: regulation of cell migration and invasion by miR-23b via targeting of Pyk2. *PLoS One*, 7, e39818.
[49] Moore, LM; Zhang, W. (2010). Targeting miR-21 in glioma: a small RNA with big potential. *Expert Opin Ther Targets*, 14, 1247-1257.

[50] Mizoguchi, M; Guan, Y; Yoshimoto, K; Hata, N; Amano, T; et al. (2012). MicroRNAs in Human Malignant Gliomas. *J Oncol*, 2012, 732874.

[51] Zhang, Y; Dutta, A; Abounader, R. (2012). The role of microRNAs in glioma initiation and progression. *Front Biosci*, 17, 700-712.

[52] Ujifuku, K; Mitsutake, N; Takakura, S; Matsuse, M; Saenko, V; et al. (2010). miR-195, miR-455-3p and miR-10a(*) are implicated in acquired temozolomide resistance in glioblastoma multiforme cells. *Cancer Lett*, 296, 241-248.

[53] Slaby, O; Lakomy, R; Fadrus, P; Hrstka, R; Kren, L; et al. (2010). MicroRNA-181 family predicts response to concomitant chemoradiotherapy with temozolomide in glioblastoma patients. *Neoplasma*, 57, 264-269.

[54] Wong, ST; Zhang, XQ; Zhuang, JT; Chan, HL; Li, CH; et al. (2012). MicroRNA-21 inhibition enhances in vitro chemosensitivity of temozolomide-resistant glioblastoma cells. *Anticancer Res*, 32, 2835-2841.

[55] Zhang, S; Wan, Y; Pan, T; Gu, X; Qian, C; et al. (2012). MicroRNA-21 inhibitor sensitizes human glioblastoma U251 stem cells to chemotherapeutic drug temozolomide. *J Mol Neurosci*, 47, 346-356.

[56] Shi, L; Chen, J; Yang, J; Pan, T; Zhang, S; et al. (2010). MiR-21 protected human glioblastoma U87MG cells from chemotherapeutic drug temozolomide induced apoptosis by decreasing Bax/Bcl-2 ratio and caspase-3 activity. *Brain Res*, 1352, 255-264.

[57] Zhang, C; Wang, G; Kang, C; Du, Y; Pu, P. (2009). [Up-regulation of p27(kip1). by miR-221/222 antisense oligonucleotides enhances the radiosensitivity of U251 glioblastoma]. *Zhonghua Yi Xue Yi Chuan Xue Za Zhi*, 26, 634-638.

[58] Chen, G; Zhu, W; Shi, D; Lv, L; Zhang, C; et al. (2010). MicroRNA-181a sensitizes human malignant glioma U87MG cells to radiation by targeting Bcl-2. *Oncol Rep*, 23, 997-1003.

[59] Lee, KM; Choi, EJ; Kim, IA. (2011). microRNA-7 increases radiosensitivity of human cancer cells with activated EGFR-associated signaling. *Radiother Oncol*, 101, 171-176.

[60] Niemoeller, OM; Niyazi, M; Corradini, S; Zehentmayr, F; Li, M; et al. (2011). MicroRNA expression profiles in human cancer cells after ionizing radiation. *Radiat Oncol*, 6, 29.

[61] Kefas, B; Godlewski, J; Comeau, L; Li, Y; Abounader, R; et al. (2008). microRNA-7 inhibits the epidermal growth factor receptor and the Akt pathway and is down-regulated in glioblastoma. *Cancer Res*, 68, 3566-3572.

[62] Li, Y; Guessous, F; Zhang, Y; Dipierro, C; Kefas, B; et al. (2009). MicroRNA-34a inhibits glioblastoma growth by targeting multiple oncogenes. *Cancer Res*, 69, 7569-7576.

[63] Guessous, F; Zhang, Y; Kofman, A; Catania, A; Li, Y; et al. (2010). microRNA-34a is tumor suppressive in brain tumors and glioma stem cells. *Cell Cycle*, 9, 1031-1036.

[64] Shi, L; Zhang, S; Feng, K; Wu, F; Wan, Y; et al. (2012). MicroRNA-125b-2 confers human glioblastoma stem cells resistance to temozolomide through the mitochondrial pathway of apoptosis. *Int J Oncol*, 40, 119-129.

[65] Chan, XH; Nama, S; Gopal, F; Rizk, P; Ramasamy, S; et al. (2012). Targeting glioma stem cells by functional inhibition of a prosurvival oncomiR-138 in malignant gliomas. *Cell Rep*, 2, 591-602.

[66] Corsten, MF; Miranda, R; Kasmieh, R; Krichevsky, AM; Weissleder, R; et al. (2007). MicroRNA-21 knockdown disrupts glioma growth in vivo and displays synergistic cytotoxicity with neural precursor cell delivered S-TRAIL in human gliomas. *Cancer Res*, 67, 8994-9000.
[67] Zhou, X; Ren, Y; Moore, L; Mei, M; You, Y; et al. (2010). Downregulation of miR-21 inhibits EGFR pathway and suppresses the growth of human glioblastoma cells independent of PTEN status. *Lab Invest*, 90, 144-155.
[68] Singh, S; Narang, AS; Mahato, RI. (2011). Subcellular fate and off-target effects of siRNA, shRNA, and miRNA. *Pharm Res*, 28, 2996-3015.
[69] Brannan, CI; Dees, EC; Ingram, RS; Tilghman, SM. (1990). The product of the H19 gene may function as an RNA. *Mol Cell Biol*, 10, 28-36.
[70] Brown, CJ; Ballabio, A; Rupert, JL; Lafreniere, RG; Grompe, M; et al. (1991). A gene from the region of the human X inactivation centre is expressed exclusively from the inactive X chromosome. *Nature*, 349, 38-44.
[71] Lee, RC; Feinbaum, RL; Ambros, V. (1993). The C. elegans heterochronic gene lin-4 encodes small RNAs with antisense complementarity to lin-14. Cell, 75, 843-854.
[72] Birney, E; Stamatoyannopoulos, JA; Dutta, A; Guigo, R; Gingeras, TR; et al. (2007). Identification and analysis of functional elements in 1% of the human genome by the ENCODE pilot project. *Nature*, 447, 799-816.
[73] Kawai, J; Shinagawa, A; Shibata, K; Yoshino, M; Itoh, M; et al. (2001). Functional annotation of a full-length mouse cDNA collection. *Nature*, 409, 685-690.
[74] Lipovich, L; Johnson, R; Lin, CY. (2010). MacroRNA underdogs in a microRNA world: evolutionary, regulatory, and biomedical significance of mammalian long non-protein-coding RNA. *Biochim Biophys Acta*, 1799, 597-615.
[75] Han, L; Zhang, K; Shi, Z; Zhang, J; Zhu, J; et al. (2012). lncRNA pro fi le of glioblastoma reveals the potential role of lncRNAs in contributing to glioblastoma pathogenesis. *Int J Oncol*, 40, 2004-2012.
[76] Zhang, X; Sun, S; Pu, JK; Tsang, AC; Lee, D; et al. (2012). Long non-coding RNA expression profiles predict clinical phenotypes in glioma. *Neurobiol Dis*, 48, 1-8.
[77] Amaral, PP; Clark, MB; Gascoigne, DK; Dinger, ME; Mattick, JS. (2011). lncRNAdb: a reference database for long noncoding RNAs. *Nucleic Acids Res*, 39, D146-151.
[78] Mercer, TR; Qureshi, IA; Gokhan, S; Dinger, ME; Li, G; et al. (2010). Long noncoding RNAs in neuronal-glial fate specification and oligodendrocyte lineage maturation. *BMC Neurosci*, 11, 14.
[79] Yang, F; Zhang, L; Huo, XS; Yuan, JH; Xu, D; et al. (2011). Long noncoding RNA high expression in hepatocellular carcinoma facilitates tumor growth through enhancer of zeste homolog 2 in humans. *Hepatology*, 54, 1679-1689.
[80] Lin, M; Pedrosa, E; Shah, A; Hrabovsky, A; Maqbool, S; et al. (2011). RNA-Seq of human neurons derived from iPS cells reveals candidate long non-coding RNAs involved in neurogenesis and neuropsychiatric disorders. *PLoS One*, 6, e23356.
[81] Gibb, EA; Vucic, EA; Enfield, KS; Stewart, GL; Lonergan, KM; et al. (2011). Human cancer long non-coding RNA transcriptomes. *PLoS One*, 6, e25915.
[82] Johnson, R. (2012). Long non-coding RNAs in Huntington's disease neurodegeneration. *Neurobiol Dis*, 46, 245-254.

[83] Michelhaugh, SK; Lipovich, L; Blythe, J; Jia, H; Kapatos, G; et al. (2011). Mining Affymetrix microarray data for long non-coding RNAs: altered expression in the nucleus accumbens of heroin abusers. *J Neurochem*, 116, 459-466.

[84] Ellis, BC; Molloy, PL; Graham, LD. (2012). CRNDE: A Long Non-Coding RNA Involved in CanceR, Neurobiology, and DEvelopment. *Front Genet*, 3, 270.

[85] Zhang, X; Zhou, Y; Mehta, KR; Danila, DC; Scolavino, S; et al. (2003). A pituitary-derived MEG3 isoform functions as a growth suppressor in tumor cells. *J Clin Endocrinol Metab*, 88, 5119-5126.

[86] Zhang, X; Rice, K; Wang, Y; Chen, W; Zhong, Y; et al. (2010). Maternally expressed gene 3 (MEG3). noncoding ribonucleic acid: isoform structure, expression, and functions. *Endocrinology*, 151, 939-947.

[87] Ji, P; Diederichs, S; Wang, W; Boing, S; Metzger, R; et al. (2003). MALAT-1, a novel noncoding RNA, and thymosin beta4 predict metastasis and survival in early-stage non-small cell lung cancer. *Oncogene*, 22, 8031-8041.

[88] Smaldone, MC; Davies, BJ. (2010). BC-819, a plasmid comprising the H19 gene regulatory sequences and diphtheria toxin A, for the potential targeted therapy of cancers. *Curr Opin Mol Ther*, 12, 607-616.

In: MicroRNA and Non-Coding RNA
Editor: James C. Johnson

ISBN: 978-1-62618-442-8

Chapter 5

Identifying miRNA Function in Innate Immunity

Jezrom B. Fordham, Afsar Raza Naqvi and Salvador Nares*
Department of Periodontology, University of North Carolina at Chapel Hill, Chapel Hill, NC, US

Abstract

miRNA regulates immunity by modulating the expression of key genes involved in leukocyte development and function. Furthermore, immunity includes innate and adaptive systems comprising myeloid-derived cells and lymphocytes which mediate nonspecific and pathogen-specific responses. Innate immunity encompasses the generation of myeloid-derived cells and their ability to mount a rapid and appropriate response to pathogen-associated danger. Here we highlight common components of research investigating the role of miRNA in innate immunity. The identification of miRNAs involved in immunity generally begins with the profiling of samples generated ex vivo or in vitro which include the cell types and immunological conditions of interest. From here a core set of methods and technologies are utilised. Various technologies exist for quantifying miRNA expression, including qRT-PCR and microarray platforms capable of profiling all known human miRNAs as listed by miRBASE. miRNA array expression data may be confirmed by qRT-PCR before in silico analysis of predicted miRNA targets, and these miRNA-target interactions validated by dual luciferase assays. Finally, the biological significance of any miRNA is dependent upon the degree of modulation it exerts combined with the importance of it target within the relevant system. A widely used approach for ascertaining this is to perform experiments in which the miRNA is constitutively expressed or knocked-down and altered function or experimental outcome tested. To date these methods have revealed several important roles for miRNA in the functioning of the immune system. As the technologies and techniques described here continue to develop so will our understanding of the function

* Corresponding author: Salvador Nares, Department of Periodontics, School of Dentistry, University of North Carolina at Chapel Hill, Chapel Hill, NC 27599-7450. Salvador_Nares@dentistry.unc.edu.

of miRNA within immunity and with our prospects of manipulating expression for therapeutic gain.

Introduction

microRNA (miRNA) expression is an integral component of the machinery that modulates gene expression at the post-transcriptional level. The biogenesis and modus operandi of miRNA has been well established and comprises the sequential activity of RNA polymerase, Drosha and Dicer enzymes to form a primary-miRNA (pri-miRNA) transcript which is converted into precursor-miRNA (pre-miRNA) and finally a mature miRNA duplex [1]. Originally the 3' strand of the duplex was believed to mediate the translational inhibition or degradation of the target mRNA; however, this canonical view has been expanded in recent times to include alternative mechanisms of pri-miRNA generation, additional processing of the pre-miRNA (notably in the let-7 family), and functionality of the 5' strand [2]. During this time new technologies have been introduced and existing technologies have been refined. These improvements have increased the range and sensitivity of miRNA detection, refined the process of messenger RNA (mRNA) target prediction and validation, and expanded the methods available for its manipulation and functional analysis.

miRNA identification and target prediction are two important aspects of miRNA research and are the cornerstone of many investigations. Perhaps the best way to describe the implementation of these and other techniques is with the use of an exemplar which will cover miRNA expression during the innate immune response. As a rule, miRNA expression appears to be more prevalent in innate than adaptive immunity. For instance, a central pillar of innate immunity is the activation of leukocytes by way of pathogen-recognition receptors (PRRs), most notably the Toll-like receptors (TLRs) and Nucleotide-binding oligomerisation domain (NOD)-like receptors, which recognise highly conserved pathogen-associated molecular patterns (PAMPs) [3]. Many PAMPs have been discovered to alter miRNA expression and numerous miRNAs are known to regulate these responses [4]. By comparison, the central mechanism of activation in adaptive immunity is the recognition of cognate antigen/MHC complexes by T cell receptors (TCR) or B cell receptors (BCR), where only a few miRNAs have been found to be involved. It's likely that such a discrepancy reflects the longer evolutionary history that miRNA-mediated gene regulation and innate-immunity have shared.

Leukocyte development is a complex process whereby haematopoietic stem cells follow branching paths which mark increasing commitment towards a final cell type. In this fashion monocytes are generated in the bone-marrow before entering the bloodstream and extravasating into tissues. Here they may differentiate into one of several possible cell types, which include macrophages (MΦs), dendritic cells (DCs) and osteoclasts (OCs). Although these cell types have different functional capacities they also share many features. Since these cells possess a common progenitor, share many morphological and functional characteristics, moreover display plasticity during immune activation, it would be reasonable to predict a role for miRNA in their differentiation and function. Here we discuss the application of miRNA technologies to help elucidate the role of miRNA in myeloid inflammatory cell development and biology.

microRNA and Myeloid Inflammatory Cell Biology

In immunity there are many examples of subtle changes having profound effects. For instance the heterozygous loss of the transcription factor PU.1 in mice, being a 2-fold reduction, alters the balance of monocyte and granulocyte production in the bone marrow. miR-155 and miR-223 have been shown to alter this balance (with the latter being regulated by PU.1) [5] Indeed, several miRNAs have been found to alter the development and function of leukocytes and innate immunity has proven to be a particularly fruitful avenue of miRNA research [6]. Confirmed targets within innate immunity include many of the receptors and ligands involved in pathogen recognition and the sensing of danger, the mechanisms of pathogen removal, and the regulation of inflammation [4].

miRNAs that function in innate immunity can be divided into two groups: those that promote pathogen recognition, inflammation and removal, as typified by miR-155, and those that inhibit these functions, as miR-146 does. miR-155 has been the focus of much investigation and many of its functions are now well-known. Its inflammatory potential is most apparent in its positive regulation of the key pro-inflammatory cytokines tumour necrosis factor-α (TNF-α), IL-6, IL-23 and the type 1 interferons (IFN) [7-10]. Alteration of its expression in mice has repercussions that include an increased susceptibility to septicaemia and a myeloproliferative disorder when over-expressed, and impaired immunity when its expression is absent [11]. miR-155 deficient mice have impaired immunity to many pathogens to the extent that immunisation is ineffective. This deficiency indicates the importance of PRR-induced NF-κB activation and the importance of NF-κB activation in upregulating miR-155 expression, although TNF-α may also induce miR-155 expression by way of AP-1 [12]. miR-155 expression also inhibits the effectiveness of anti-inflammatory signalling by downregulating receptors such as IL-13 and intracellular-signalling molecules such as SHIP1 and SOCS1 [13].

Opposing these effects are the anti-inflammatory miRNAs like miR-146 which is also upregulated by NF-κB activation and TNF-α signalling. Interestingly miR-146 upregulation appears to be biased towards PRRs that recognise bacterial PAMPs, indicating that secondary signals to NF-κB activation influence its expression. miR-146 also acts as a negative regulator of both its own expression and the activation of innate, pro-inflammatory responses, which it does by inhibiting signalling components of NF-κB activation (including TRAF6 and IRAK1) [14]. Another miRNA, miR-21, acts in a similar fashion and is mainly upregulated by the TLR4/NF-κB pathway. Here miR-21 targets PDCD4, another activator of NF-κB. The aforementioned miRNAs have been widely investigated but only represent a fraction of the PRR-induced miRNAs the functions of which remain to be discovered.

Although the regulatory functions of miRNA appear to be more prevalent in innate than adaptive immunity, there are some significant exceptions. Take central tolerance, where positive and negative selection in the thymus mediates the removal of highly auto-reactive immature T cells. Here miR-181a contributes to the regulation of this process. In mice deficient for miR-181a the threshold for the negative selection of immature T cells is higher resulting in an increased number of autoreactive T cells being positively selected for [15]. In the periphery T helper (Th) cells are important for coordinating adaptive responses, with different Th subsets being tailored to the removal of different types of pathogen. miR-155 promotes the differentiation of two of these subsets: Th1 and Th17,which support anti-microbial and cell-mediated immune responses. Without miR-155 expression Th2 responses which mainly support B cell maturation and aid in the clearance of extracellular pathogens are dominant [11].

Application of miRNA Technologies to Investigate Myeloid Cell Differentiation and Function

Identification of the miRNAs expressed during myeloid cell differentiation would begin with the acquisition of monocytes. In this section we will provide examples using primary human cells although the techniques described herein also apply to other cell types. $CD14^+$ monocytes are easily isolated from peripheral blood mononuclear cells (PBMCs) obtained from human donors by density gradient sedimentation (Ficoll-Hypaque) and magnetic-activated cell sorting (MACS). Greater numbers of cells can be obtained using counterflow centrifugal elutriation in the order of 1×10^8-2×10^9 from 300 milliliters of blood obtained by leukapheresis [16]. In the absence of leukapheresis, approximately 2×10^8-4×10^8 monocytes can still be recovered from 1 unit of whole blood using this procedure. These methods typically results in >95% purity, based on flow cytometric analysis of CD14 expression. miRNA profiling of monocytes would be considered the baseline from which differentiation-induced changes would be calculated. For the differentiation of MΦs, macrophage-colony stimulating factor (M-CSF) or granulocyte-macrophage colony stimulating factor (GM-CSF) may be added to the culture medium, while interleukin-4 (IL-4) and GM-CSF is required to induce the differentiation of DCs [17, 18]. Osteoclasts are generated from monocytes with a combination of soluble (sRANKL) and M-CSF. The time taken for full differentiation can be assessed by characteristic changes in morphology (Figure 1A) and flow cytometric analysis of cell-type specific markers. Four useful markers are CD14, CDw93, CD1a and CD209 and their differential expression by monocytes, MΦs and DCs is shown in Table 1. The extreme size and adhesive properties of osteoclasts requires methods other than flow cytometry for their identification, which can include size, number of nuclei and tartrate-resistant acid phosphatase (TRAP) expression (Figure 1B).

Flow cytometry may indicate that complete differentiation of MΦs and DCs requires seven days of culture while OC differentiation may take up to six days with functionality (i.e., bone degradation) increasing over the next seven to fourteen days. As such, analysis of miRNA expression over this extended time requires careful consideration of the kinetics of miRNA expression. Like mRNA, changes in miRNA expression can be transient or stable and are influenced by numerous factors including the rate of transcription, the stability of the transcript and mechanisms of targeted-degradation or alteration. Several factors that regulate miRNA expression and functional outcome have recently been identified. The classical innate response to viral infection: interferon production and signalling, can down-regulate Dicer expression to limit the rate of mature miRNA production. While at a more local level the transcription factors Smad and p53, and the mRNA-decay promoting factor KSRP, appear to regulate miRNA synthesis and may specifically contribute to increased miR-155 expression during the inflammatory response. Intriguingly inflammation can also induce a form of miRNA-editing by activating the adenosine deaminase Adar. By converting adenosines to inosines Adar can alter the target specificity of a miRNA, however the biological relevance of this editing remains to be seen.

miRNA expression profiling begins with the isolation of total RNA and various protocols are available [19]. It is important that any isolation method be inclusive of small RNAs (20-50nt) while the concomitant capture of larger RNA molecules is useful for analysis of mRNA target species from the same sample. One widely used method is the chemical extraction protocol which utilises chaotropic salts (e.g., QIAzol [Qiagen] & Trizol [Invitrogen]) in conjunction with silica-based columns. Kits containing these reagents are available from a number of manufacturers and provide a rapid and efficient means of RNA isolation.

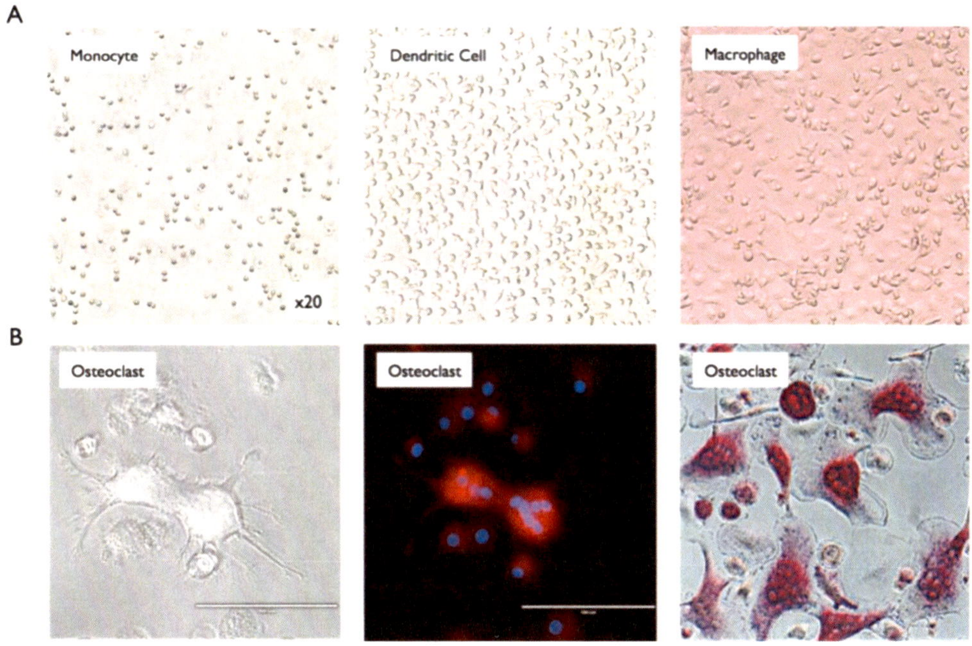

Figure1. A) Characteristic morphology of monocytes, macrophages and dendritic cells. B) Characteristic morphology of differential osteoclasts (left panel), with staining for nuclei (blue) and lysosomes (red) (Centre). Osteoclast expression of TRAP (right panel).

Table 1. Expression of differentiation markers by monocytes, macrophages, dendritic cells and osteoclasts

Cell Type	Differentiation Marker						
	CD14	CDw93	CDla	CD209	Multi-nucleated	TRAP	Bone Degradation
Monocyte	+	+	-	-	-	-	-
Macrophage	+	-	-	-	-	-	-
Dendritic Cell	-	-	+	+	-	-	-
Osteoclast	-	-	-	-	+	+	+

Assessment of RNA quantity, purity and integrity is essential as degraded RNA and contaminants will have a negative affect on the accuracy and reproducibility of results derived from downstream applications such as quantitative real time-polymerase chain reaction (qRT-PCR) assays or miRNA microarrays [20]. Measurement of absorption by ultraviolet spectrophotometry using precision instruments such as the Nanodrop (Agilent) provides a good estimate of concentration (260nm) and purity (260nm/280nm and 260nm/230nm ratios). For assessment of RNA integrity electrophoresis-based methods such as Agilent's Bioanalyzer or the Bio-Rad's Experion are useful. RNA Integrity Number (RIN), with a score ≤ 5 indicates poor quality RNA while a score of > 6 indicates reasonably good quality RNA for further study [21]. Profiling of single or small numbers of miRNAs or validation microarray data can be achieved using qRT-PCR or Northern blotting. Northern blotting requires larger quantities of RNA while qRT-PCR requires a much smaller sample and has a wider dynamic range of quantification. For example, qRT-PCR can be routinely used to monitor expression levels of miR-155 and miR-146 in myeloid cells after challenge with lipopolysaccharide (LPS). TaqMan or SYBR-green based qRT-PCR assays are commercially available some of which can be bought as low-density 'PCR arrays' in 96- or 394-well plate format tailored to known biological functions and pathways (e.g. inflammation, cell differentiation, cancer, etc.). Northern blotting has the advantage in that interrogation of pre-mRNA levels (presenting as a higher molecular weight band) identification of 3' and 5' length variants, and confirm the molecular weight of a miRNA is possible. This was one of the first technologies to be used for large-scale analysis of miRNA expression and today these are offered as technical services (Exiqon and Affymetrics).

The rapid expansion of the field has been supported by advances in miRNA profiling technology, with modern platforms now capable of analysing the expression of many thousands of miRNAs and with greater sensitivity than ever before. While early efforts in profiling faced technical difficulties due to the short nucleotide length and variable melting temperatures of miRNAs, plus the high sequence similarity that exists between family members, new technologies have overcome these issues. For example, the use of locked nucleotide RNA (LNA) technology (locked referring to the constraint of the ribose ring by the addition of a methylene bridge between the O2' and C4' atoms) has improved microarray hybridization technology [21]. As our ability to detect miRNA expression has deepened so the known pool of miRNAs expressed by each species has widened. Novel miRNA sequences are still being found for many species, including viruses, with approaches such as genome wide deep-sequencing proving useful, and the central repository of sequences. Indeed, miRBase now contains over 20,000 mature mRNA sequences (miRBase v18.0, accessed 10 December 2012) spanning close to 200 different species.

For each miRNA its target mRNA can be predicted in silico using algorithms such as Targetscan, miRWalk, MiRanda and others [22]. These algorithms are regularly updated as our understanding of miRNA biology has progressed. Early predictions were based purely on miRNA-mRNA sequence complementarity with approximately 10% of predicted targets translating into true targets. This has since been adapted to factor in intergenic conservation between species, and while this increases true target identification to 60% it also omits approximately 40% of the valid targets [23, 24]. Current algorithms, such as miRanda, will cross-reference a miRNA sequence with known 3' UTRs (according to the Ensembl genome annotation database) and assign a target score still based upon the degree of sequence complementarity but adjusted according to the complementarity at the seed region, the presence of multiple targets within the 3' UTR, and the degree of intergenic conservation.

Also, the recently introduced Nanostring nCounter (which is still based on novel hybridization technology) uses an innovative approach in which sequence-specific capture probes tailored for miRNAs of interest are used in the creation of a multiplexed probe library [25] The main advantage of the Nanostring is its ability to discriminate between similar variants of a miRNA with a high degree of accuracy. High-throughput sequencing, termed RNA-seq, has emerged as an application of next-generation sequencing (NGS) technologies and can also be used for miRNA profiling experiments [26]. This approach uses a small RNA cDNA library that has been generated from the sample for the 'massively parallel' sequencing of each cDNA. Bioinformatic analysis of the sequence reads then identifies miRNA and digital methods are applied for relative quantification.

Data analysis begins once miRNA profiling is complete. Visualization of global profiling data can be presented as a heatmap where the relative expression of each miRNA is presented on a dual-coloured spectrum (Figure 2A), or alternatively as a volcano plot where miRNA expression is plotted against p-value. For a comprehensive description of the analytical approaches to miRNA profiling data the reader is referred to recent reviews on the subject [24, 27]. These provide a convenient visual representation of the most significantly differentially expressed miRNAs. However for the differentiation-induced changes in MΦs and DCs, miRNA expression would be calculated for each time-point using monocyte expression as a baseline. From this analysis, baseline miRNAs that display statistically significant changes compared to monocytes would be candidates for functional analysis. In our example the number of differentially expressed miRNAs was found to increase over time for both MΦs and DCs (Figure 2B).

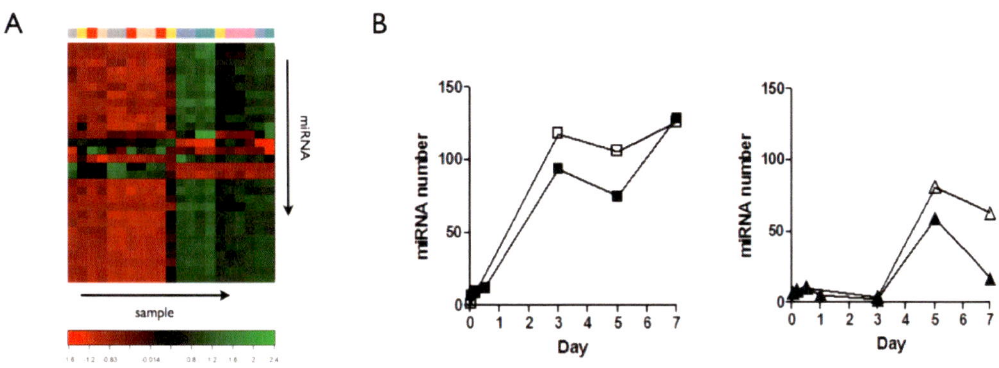

Figure 2. A) Heat-map representation of miRNA expression. On a spectrum from red (denoting downregulation) to green (upregulation) the magnitude of the fold-change in individual miRNA expression is indicated by intensity. B) miRNA profiling of monocytes as they differentiate into macrophages (left panel) and dendritic cells (right panel).

The selection of candidate miRNAs is based upon two central criteria: the fold change in expression (coupled with statistical significance) and functional the importance of the predicted mRNA targets. The greater the fold change the more likely it is to alter mRNA expression, with the caveat that basal levels of expression in different tissues or cells is also a factor. While the average copy number for an individual miRNA is ~500, its expression across different cell types can be over a range of four orders of magnitude [28]. For instance, a 2-fold upregulation of a given miRNA may be comparable to a 4-fold upregulation of an alternate miRNA species if its basal expression is 2-fold greater than the former.

One example is the activation of B cells with LPS which results in the upregulation of miR-155 and -146a. Although this induction of miR-155 expression is greater than that of miR-146a, the higher basal level of miR-146a results in comparable levels. This is an important consideration as miRNA expression only needs to reduce mRNA copy numbers below threshold levels. For while protein production is sensitive to changes in mRNA expression it is highly repressed when it drops below threshold levels. Moreover, modulation of gene expression by miRNA is not solely based on the ratio of miRNA to mRNA copy number, but also includes the strength of the miRNA-mRNA interaction. This implies that a small increase in the expression of a highly inhibitory miRNA may be equal or greater than a larger increase in a low-inhibitory miRNA. Work of this nature has clarified how certain miRNAs can function as molecular switches (e.g., miRNA lin-4 in *C. elegans)* as well as fine-tuners of gene expression [29]. As such, the use of highest fold-change as the primary consideration for selection may ignore miRNAs with smaller expression changes but of significant biological importance. Further information on the challenges of miRNA target prediction can be found in the recent review by Barbato [27].

Initial confirmation of miRNA-mRNA interactions can be performed using reporter assays. Options include the dual-luciferase reporter system, and recently developed methods such as the introduction of biotin-tagged miRNA duplexes into cells followed by the isolation of miRNA-mRNA complexes from cell lysates using streptavidin beads [30]. For conventional luciferase reporter assays, cloning of the target 3'UTR downstream of the luciferase gene followed by transfection of the vectors and miRNA mimic into a cell line such as the HEK293 is commonplace. A decrease in the luciferase readout relative to controls would indicate potential miRNA modulation of the target gene. At this point further confirmation would require analysis of protein expression.

Further confirmation that a particular miRNA alters protein expression and hence a biological function, can be achieved by overexpression or inhibition studies. miRNA inhibition involves the introduction of an antisense oligonucleotide, while miRNA mimics act as mature miRNAs. The use of mimics raises two points that do not apply to the use of inhibitors. First, the concentration of the mimic may be much higher than present under normal conditions, and while this may confirm miRNA-mRNA interaction it does not confirm its relevance to biological function [31]. Second, the complementary strand of the mimic may cause off-target effects, although careful titration of the mimic may limit or exclude this possibility [32].

Various methods exist for the transfer of oligonucleotides into cells. Established techniques that are efficient for transferring short nucleotide sequences are also suitable for experiments that utilise miRNA mimics and inhibitors. The choice of which method to use usually depends on the degree of resistance to transfection, whether stable or transient transfection is required, and the degree of activation or cell death induced by the method. For long-term expression viral vectors are the most useful and this method possesses distinct advantages. Cells transduced in this way display stable expression and if coupled to response elements, expression can be induced or repressed through promoter activity. Inducible-expression using viral vectors also implies that rescue of phenotype studies can be performed using a single vector rather than requiring multiple transfections or overexpression/knockdown genetic manipulations. Potential drawbacks to viral vectors include the induction of type 1 interferons and apoptotic responses as well as the resistance of certain cell types, such as monocytes, to infection. Still, where large numbers of cells are available for transduction the application of fluorescence activated cell-sorting (FACS) may circumvent these issues.

Other methods include electroporation and chemical transfection. Chemical methods transfect cells transiently, although this may still constitute four days or more, and have the added benefit of being highly efficient for the transduction of many cell types. Electroporation can be particularly useful for transient transfection of cell types that are challenging to transfect by other methods but require specialized equipment. In all cases transfected cells are identifiable if fluorescence-labeled oligonucleotides are used. In MΦ and DC, we employ HiPerfect transfection reagent in Opti-MEM medium (Gibco) beginning two days after monocyte differentiation. Here an estimate of transfection efficacy is observed by fluorescence microscopy (Figure 3A) or quantified by flow cytometry (Figure 3B). Alternatively this can be measured using qRT-PCR (Figure 3C).

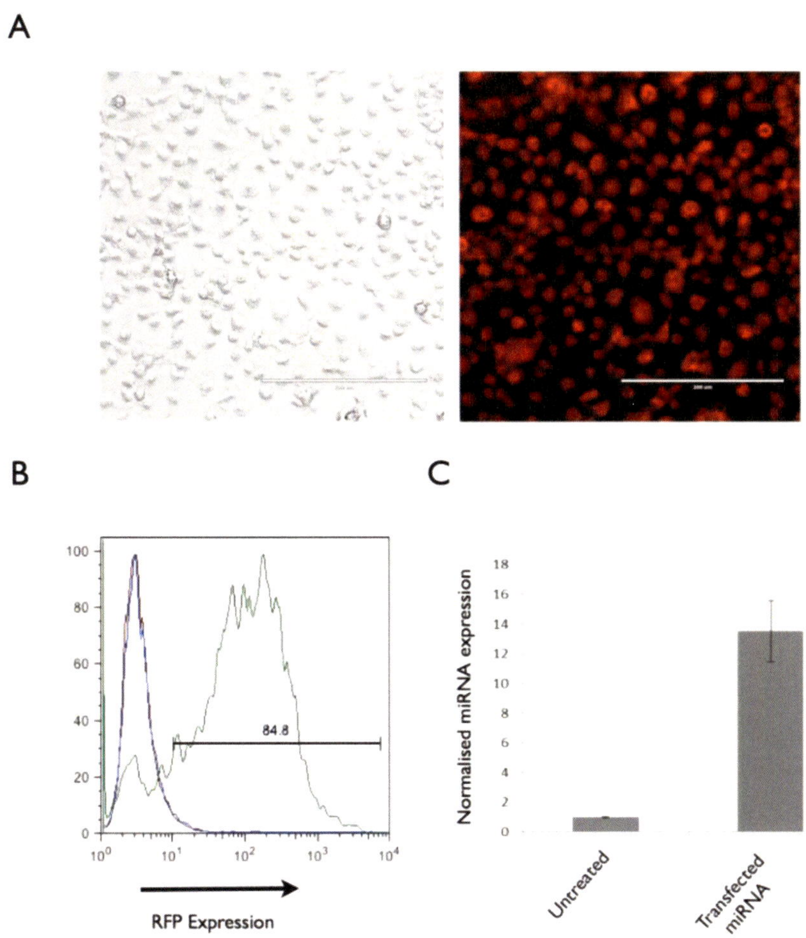

Figure 3. A) Light (left) and Fluorescent microscopy (right) images of macrophages transfected with a red fluorescent protein (RFP)-labelled oligonucleotide. B) Quantification of transfection efficiency by flow cytometry. C) Quantification of transfection efficiency by qRT-PCR.

Oligonucleotide with scrambled miRNA sequences have been used as transfection controls but were often plagued with off-target effects. These have been largely superseded by alternatives such as species-specific miRNA sequences. For example, selection of sequences identified only in *C. elegans* (such as Qiagen) are useful as controls for human studies as there are no predicted targets found in humans [32]. Opposing or negligible effect should be realized when using mimics and inhibitors, depending upon whether basal level of expression is sufficient for regulation or not.

The human genome encodes ~30,000 mRNAs and up to a half of these may be regulated by the ~1000 miRNAs found in the genome. Understanding this regulation requires the analysis of networks formed by a single miRNA which can target hundreds of mRNAs. Identifying the networks formed by miRNA-mRNA interactions is challenging. Much more is known about the networks formed by protein-protein interactions and how they alter cell function but these are also complex. The challenge presented in linking changes in the expression of multiple miRNAs to changes in multiple proteins, and predicting the physiological process most likely to be regulated is considerable. Faced with this complexity, the myopic view of a single miRNA-mRNA-protein interaction is the easiest to see.

For a predicted miRNA target there should be a consensus among several predictive algorithms, with each one providing a high target score. The target should also be expressed by the tissue or cell type in question and be of relevance to the experimental conditions in question. Identification of a single miRNA-mRNA interaction resulting in significant downregulation of protein is the simplest and most commonly published observation. In reality that particular miRNA may (due to partial mismatching and the bias of 3' UTR targeting to the much shorter seed region) regulate hundreds of mRNAs, inhibiting some and degrading others [23]. Individually none of these may not significantly affect protein expression or cellular function, yet when altered in unison their role may become apparent. The other limitation of focusing on individual miRNA-mRNA interactions is that it does not take into account the ability of multiple miRNAs, each possessing a negligible regulatory capacity, to target a single mRNA to achieve inhibition [23]. The future of miRNA research will include the mapping of functional networks from single miRNA to multiple mRNA- and multiple miRNA to single mRNA-interactions. Knowledge of these networks will be particularly important in the context of therapeutic manipulation where changes in one miRNA target may be advantageous while others may be deleterious to the host. The increasing application of analytical programs like Uniquity and mirConnX which can cross-reference changes in miRNA, mRNA and protein expression to generate networks should accelerate this process [33].

So how has this technology been harnessed to further our understanding of miRNA function in immunity? First it should be noted that the reputation of miRNA as mainly being fine-tuners holds true for immunity. Still, to extend the analogy a small turn of the dial can mean the difference between a coherent signal and white noise. In immunity this may mean the difference between an appropriate response to pathogen-associated danger or a dysregulated response leading to chronic inflammation, immunopathology or autoimmunity.

To return to innate immunity and our exemplar experiment, the profiling is likely to reveal numerous miRNAs differentially expressed over time and between cell types. Our three cell types: monocytes, MΦs and DCs, share many functions while also possessing adaptations that better allow them to fulfill their distinct roles in immunity. It is reasonable to hypothesise that when changes in an individual miRNA are present in multiple cell types it may be contribute to a shared function. By contrast, cell-type specific changes are more likely to contribute to functions that are not shared. Here phagocytosis is a shared function but the downstream responses in each cell type are different. In monocytes the main outcome of phagocytosis is the production of pro-inflammatory cytokines (also present in MΦs and DCs). The phagocytic function in MΦs is important for anti-microbial responses and in the restoration of homeostasis following tissue inflammation [34, 35]. In DCs the most important outcome of phagocytosis is antigen presentation and the initiation of adaptive immunity [36]. In silico analysis may predict that a miRNA interacts with many targets that are known to be involved in phagocytosis. To investigate the functional capacity of miRNA to augment or inhibit a particular biological function in this case phagocytosis, cells can be transfected with synthetic oligonucleotides and their phagocytic activity assayed using fluorescent-labelled bioparticles, with quantification by fluorescent microscopy (Figure 4A) or flow cytometry (Figure 4B).

Identification of a cellular function that is regulated by a miRNA is an important step in the discovery process, but just one of many. To complete the picture the mRNAs that are downregulated by the miRNA must be confirmed, as well as those that are responsible for the alteration in cellular function. The recent introduction of techniques for analysing miRNAs that are bound to their targets (for example cross-linking with immunoprecipitation [CLIP]) will accelerate the validation of true miRNA-target interactions [37].

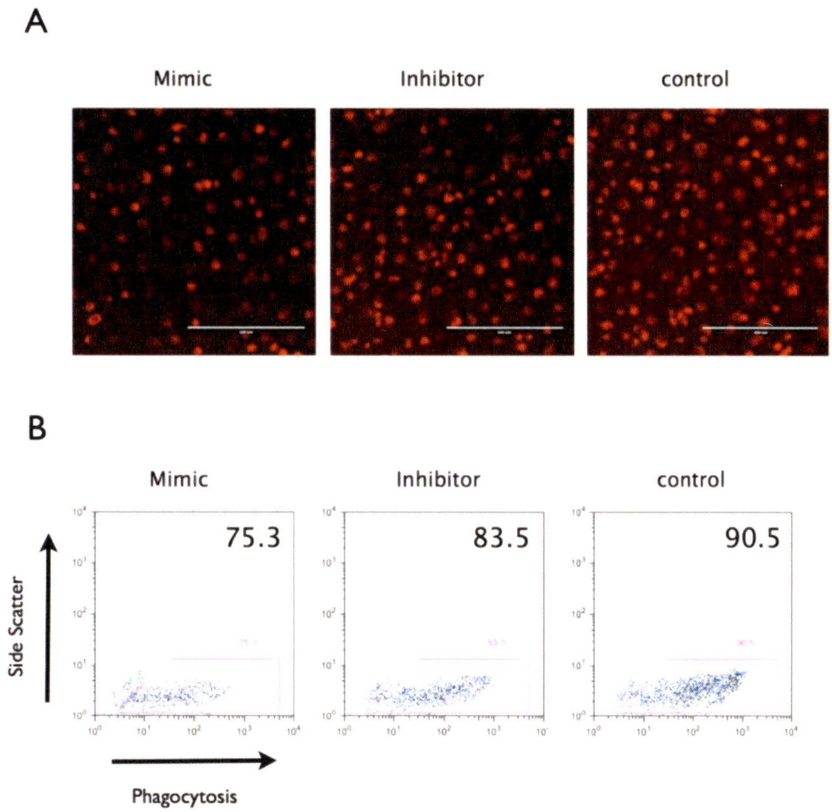

Figure 4. A) Phagocytic uptake of red fluorescent protein (RFP)-labelled bioparticles by microphages transfected with a synthetic miRNA oligonucleotide (left panel), anti-sense oligonucleotide (centre), or negative control oligonucleotide derived from C.elegans (right). B) Quantification of phagocytosis by flow cytometry.

Where immunity is the main focus the role of the miRNA may be further explored using fresh or archived human biopsy samples as well as transgenic mice and appropriate models of immunity and disease. The accessibility and sensitivity of detection of miRNA makes them attractive candidates as disease biomarkers, and several miRNAs have already been found whose expression is specific for certain types of cancer [38]. Looking to the future miRNA expression profiling may be used in the prognosis of disease or in choosing the best type of therapy for different pathologies. These advances will mark important stepping stones in the translation of miRNA research to miRNA-based clinical therapy.

Conclusion

miRNA has emerged as an information-rich component of the cells regulatory machinery and one that has an impact on the most fundamental processes of biology. Its expression alters cell development (the beginning), function (the middle) and apoptosis (the end). Our understanding of the role miRNA expression plays in immunity has also expanded, and it is now evident that miRNA contributes to the recognition, response, and removal of non-self from self (survival). Equally its aberrant expression can contribute to disease susceptibility and morbidity, including autoimmunity and immunopathology, and ultimately to mortality. As we continue to uncover the pathways between miRNA and protein expression, and to further our knowledge of their cause and effect, so we also build a roadmap towards their therapeutic manipulation. If miRNA-based therapy can deliver on its potential to improve human health then one hopes that the road is approximately 22 nucleotides long.

Acknowledgement

This study was funded by the U.S. National Institutes of Health (DE021052 [SN]).

References

[1] Krol, J., I. Loedige, and W. Filipowicz, The widespread regulation of microRNA biogenesis, function and decay. *Nat Rev Genet*, 2010. 11(9): p. 597-610.

[2] Davis-Dusenbery, B.N. and A. Hata, Mechanisms of control of microRNA biogenesis. *Journal of Biochemistry*, 2010. 148(4): p. 381-392.

[3] Kawai, T. and S. Akira, The roles of TLRs, RLRs and NLRs in pathogen recognition ARTICLE. *International Immunology*, 2009. 21(4): p. 317-337.

[4] O'Neill, L.A., F.J. Sheedy, and C.E. McCoy, MicroRNAs: the fine-tuners of Toll-like receptor signalling. *Nat Rev Immunol*, 2011. 11(3): p. 163-175.

[5] Johnnidis, J.B., et al., Regulation of progenitor cell proliferation and granulocyte function by microRNA-223. *Nature*, 2008. 451(7182): p. 1125-1129.

[6] Fontana, L., et al., MicroRNAs 17-5p-20a-106a control monocytopoiesis through AML1 targeting and M-CSF receptor upregulation. *Nat Cell Biol*, 2007. 9(7): p. 775-787.

[7] Jing, Q., et al., Involvement of MicroRNA in AU-Rich Element-Mediated mRNA Instability. *Cell*, 2005. 120(5): p. 623-634.
[8] O'Connell, R.M., et al., MicroRNA-155 Promotes Autoimmune Inflammation by Enhancing Inflammatory T Cell Development. *Immunity*, 2010. 33(4): p. 607-619.
[9] Tili, E., et al., Modulation of miR-155 and miR-125b Levels following Lipopolysaccharide/TNF-α Stimulation and Their Possible Roles in Regulating the Response to Endotoxin Shock. *The Journal of Immunology*, 2007. 179(8): p. 5082-5089.
[10] Wang, P., et al., Inducible microRNA-155 Feedback Promotes Type I IFN Signaling in Antiviral Innate Immunity by Targeting Suppressor of Cytokine Signaling 1. *The Journal of Immunology*, 2010. 185(10): p. 6226-6233.
[11] Rodriguez, A., et al., Requirement of bic/microRNA-155 for Normal Immune Function. *Science*, 2007. 316(5824): p. 608-611.
[12] Tserel, L., et al., MicroRNA Expression Profiles of Human Blood Monocyte-derived Dendritic Cells and Macrophages Reveal miR-511 as Putative Positive Regulator of Toll-like Receptor 4. *Journal of Biological Chemistry*, 2011. 286(30): p. 26487-26495.
[13] Martinez-Nunez, R.T., F. Louafi, and T. Sanchez-Elsner, The Interleukin 13 (IL-13) Pathway in Human Macrophages Is Modulated by MicroRNA-155 via Direct Targeting of Interleukin 13 Receptor α1 (IL13Rα1). *Journal of Biological Chemistry*, 2011. 286(3): p. 1786-1794.
[14] Zhao, J.L., et al., NF-κB dysregulation in microRNA-146a–deficient mice drives the development of myeloid malignancies. *Proceedings of the National Academy of Sciences*, 2011. 108(22): p. 9184-9189.
[15] Li, Q.-J., et al., miR-181a Is an Intrinsic Modulator of T Cell Sensitivity and Selection. *Cell*, 2007. 129(1): p. 147-161.
[16] Nares S, W.S., *Monocytes and macrophages. Measuring Immunity: Basic Science and Clinical Practice*, 2005. 1st Edition: p. 299-311.
[17] Fleetwood, A.J., et al., Granulocyte-Macrophage Colony-Stimulating Factor (CSF) and Macrophage CSF-Dependent Macrophage Phenotypes Display Differences in Cytokine Profiles and Transcription Factor Activities: Implications for CSF Blockade in Inflammation. *The Journal of Immunology*, 2007. 178(8): p. 5245-5252.
[18] Jansen, B., et al., MicroRNA genes preferentially expressed in dendritic cells contain sites for conserved transcription factor binding motifs in their promoters. *BMC Genomics*, 2011. 12(1): p. 330.
[19] Accerbi, M., et al., *Methods for Isolation of Total RNA to Recover miRNAs and Other Small RNAs from Diverse Species*, B.C. Meyers and P.J. Green, Editors. 2010, Humana Press. p. 31-50.
[20] Becker, C., et al., mRNA and microRNA quality control for RT-qPCR analysis. *Methods*, 2010. 50(4): p. 237-243.
[21] Castoldi, M., et al., miChip: A microarray platform for expression profiling of microRNAs based on locked nucleic acid (LNA) oligonucleotide capture probes. *Methods*, 2007. 43(2): p. 146-152.
[22] Thomas, M., J. Lieberman, and A. Lal, Desperately seeking microRNA targets. *Nat Struct Mol Biol*, 2010. 17(10): p. 1169-1174.
[23] Selbach, M., et al., Widespread changes in protein synthesis induced by microRNAs. *Nature*, 2008. 455(7209): p. 58-63.

[24] Wu, S., et al., Multiple microRNAs modulate p21Cip1/Waf1 expression by directly targeting its 3[prime] untranslated region. *Oncogene*, 2010. 29(15): p. 2302-2308.
[25] Geiss, G.K., et al., Direct multiplexed measurement of gene expression with color-coded probe pairs. *Nat Biotech*, 2008. 26(3): p. 317-325.
[26] Metzker, M.L., Sequencing technologies [mdash] the next generation. *Nat Rev Genet*, 2010. 11(1): p. 31-46.
[27] Barbato, C., et al., Computational Challenges in miRNA Target Predictions: To Be or Not to Be a True Target? *Journal of Biomedicine and Biotechnology*, 2009. 2009.
[28] Lim, L.P., et al., The microRNAs of Caenorhabditis elegans. *Genes & Development*, 2003. 17(8): p. 991-1008.
[29] Mukherji, S., et al., MicroRNAs can generate thresholds in target gene expression. *Nat Genet*, 2011. 43(9): p. 854-859.
[30] Thomson, D.W., C.P. Bracken, and G.J. Goodall, Experimental strategies for microRNA target identification. *Nucleic Acids Research*, 2011.
[31] Bracken, C.P., et al., A Double-Negative Feedback Loop between ZEB1-SIP1 and the microRNA-200 Family Regulates Epithelial-Mesenchymal Transition. *Cancer Research*, 2008. 68(19): p. 7846-7854.
[32] Caffrey, D.R., et al., siRNA Off-Target Effects Can Be Reduced at Concentrations That Match Their Individual Potency. *PLoS ONE*, 2011. 6(7): p. e21503.
[33] Huang, G.T., C. Athanassiou, and P.V. Benos, mirConnX: condition-specific mRNA-microRNA network integrator. *Nucleic Acids Res*, 2011. 39(Web Server issue): p. W416-23.
[34] Underhill, D. and A. Ozinsky, Phagocytosis of microbes: Complexity in action. *Annual Review of Immunology*, 2002. 20: p. 825-852.
[35] Savill, J., Apoptosis in resolution of inflammation. *Journal of Leukocyte Biology*, 1997. 61(4): p. 375-80.
[36] Mellman, I. and R.M. Steinman, Dendritic Cells: Specialized and Regulated Antigen Processing Machines. *Cell*, 2001. 106(3): p. 255-258.
[37] Ule, J., et al., CLIP Identifies Nova-Regulated RNA Networks in the Brain. *Science*, 2003. 302(5648): p. 1212-1215.
[38] Boeri, M., et al., MicroRNA signatures in tissues and plasma predict development and prognosis of computed tomography detected lung cancer. *Proceedings of the National Academy of Sciences*, 2011. 108(9): p. 3713-3718.
[39] Huang, J.C., et al., Using expression profiling data to identify human microRNA targets. *Nat Meth*, 2007. 4(12): p. 1045-1049.

In: MicroRNA and Non-Coding RNA
Editor: James C. Johnson

ISBN: 978-1-62618-442-8

Chapter 6

Multiple IsomiRs and Diversity of miRNA Sequences Unveil Evolutionary Roles and Functional Relationships across Animals

Li Guo[1*], Feng Chen[1] and Zuhong Lu[2]
[1]Department of Epidemiology and Biostatistics and Ministry of Education Key Lab for Modern Toxicology, School of Public Health, Nanjing Medical University, Nanjing, China
[2]Department of Biomedical Engineering, College of Engineering, Peking University, Beijing, China

Abstract

As crucial negative regulatory molecules, small non-coding RNAs, microRNAs (miRNAs), have been widely studied, especially for their functions and evolutions. Multiple miRNA variants, termed as isomiRs, are recently detected and identified from a given miRNA locus due to alterative and imprecise cleavage of Drosha and Dicer. Compared to the canonical miRNA sequence, these isomiRs may be found various 5' and/or 3' ends (various 3' ends are more popular), 3' additional non-template nucleotides, 5' and seed modification events. Similarly, miRNA sequences in different animal species also show diversity as well as species specificity. The miRNA sequences in different animal species may be involved in different length distributions, 5' and/or 3' ends, transition/transversion and insertion/deletion. From a given miRNA locus, internal multiple isomiRs and various miRNA sequences across species are quite similar and have close relationships. Indeed, they also implicate potential functional and evolutionary relationships. By applying high-throughput sequencing technique, multiple isomiRs from

* E-mail addresses: LG: gl8008@163.com, lguo@njmu.edu.cn; FC: fengchen@njmu.edu.cn; ZHL: zhlu@pku.edu.cn.

a given miRNA locus can be identified with different sequence counts, length distributions and various expression patterns, although isomiR repertoire and their expression patterns always are stable across different tissues, samples and even animal species. Most isomiRs were 3' isomiRs which had the same seed sequences, while 5' isomiRs were found to have new seed sequences via the phenomenon of "seed shifting". Many miRNAs were detected a set of isomiRs, in which 1-3 were abundant isomiRs characterized by a narrow length distribution (20-23 nt) due to dominant cleavage. Indeed, specific miRNA sequences have been identified in specific animal species through classical experimental validation. The annotated miRNA sequence is always derived from the most abundantly expressed isomiR, which is prone to be detected by traditional detection. Therefore, we speculate that the most dominant isomiR is not randomly selected, but is strictly regulated by complex regulatory network from selection and evolutionary pressures. Various miRNA sequences can unveil dynamic evolutionary processes across different animal species (cross-species) and across different miRNAs (within species), especially simultaneously involved in the diversity of pre-miRNA and pri-miRNA sequences. Furthermore, more miRNA* sequences, ever as passenger strands of miRNA sequences, are identified as active and functional miRNA sequences. These potential functional miRNA* sequences and potential more functional isomiRs (such as 5' isomiRs with novel seed sequences by changing 5' ends) will complicate miRNA-mRNA and miRNA-miRNA interaction, and enrich coding-non-coding RNA regulatory network.

Keywords: microRNA (miRNA); isomiRs; evolution; function

Abbreviations

bfl	Branchiostoma floridae
bta	Bos taurus
cap	Capitella sp.1
cfa	Canis familiaris
cgr	Cricetulus griseuss
dre	Danio rerio
gga	Gallus gallus
has	Homo sapiens
mdo	Monodelphis domestica
mml	Macaca mulatta
mmu	Mus musculus
oan	Ornithorhynchus anatinus
rno	Rattus norvegicus
sko	Saccoglossus kowalevskii
ssc	Sus scrofa
xtr	Xenopus tropi

Introduction

MicroRNAs (miRNAs) are small non-coding RNAs that are involved in post-transcriptional gene regulation by targeting mRNAs for translational repression, cleavage or destabilization in animals and plants [1,2]. In animals, miRNAs are derived in a two-step process, in which the nuclear-localized RNaseIII Drosha defines one end of the miRNA-miRNA* duplex and releases miRNA precursors (pre-miRNA hairpins) [3].

These miRNA precursors are then exported to the cytoplasm and released as short double-stranded RNA duplexes following a second cleavage by Dicer1 [4]. Based on the thermodynamic stability of each end of this duplex, one of the strands is thought to be a biologically active miRNA, and the other is considered as inactive strand called miRNA* (star) or passenger strand [5]. However, recent studies showed that some miRNA* sequences express abundantly as guide miRNAs and contribute to regulation network [6-9]. The imprecision of Drosha or Dicer cleavage could result in the production of a set of miRNA-miRNA* duplexes with a variety of 5' and/or 3' ends, which either alters the seed sequences or inverts the relative stabilities of the 5' and 3' ends of the duplex [10].

Many miRNAs are conserved phylogenetically across large evolutionary distances like vertebrates and fruit flies, especially their seed sequences (nucleotides 2-8) [11-13]. They are used as a guide to direct negatively post-transcriptional regulation by the binding of 5'-seeds (nucleotides 2-8) and anchor (nucleotides 13-16) with target sequences in the 3' untranslated region (UTR) of cognate mRNAs [14,15]. The small RNAs may be preferentially expressed in certain tissues, during specific stages of development, or in disease [15-17], and decisive roles for miRNAs have been described across mammalian development and human diseases. miRNA is widely studied by traditional experimental and computational methods, especially recently developed "next generation sequencing" (NGS) technologies, including Illumina sequencing (Solexa sequencing), 454 Life Sciences (Roche) and SOLiD™ System (ABI).

These technologies afford an unprecedented opportunity to discover and profile known and novel miRNAs at unprecedented sensitivity, and allow identification of hundreds of millions of short RNAs in a sample in a single run. Based on current sequencing technologies, multiple isomiRs, the population of variants of known miRNAs, were reported in animals and plants [18-31]. Variation at the 3' end is generally much more common than at the 5' end [20, 21, 32-34], which appears to derive from variability in Dicer and Drosha cleavage sites.

The most abundant sequence in these various isomiRs may be not correspond exactly to the current miRBase reference sequence, and relative abundance of isomiRs may vary across tissues [32]. Although the biological consequence of isomiRs remains to be determined, miRNA variants might influence the miRNA half-life, subcellular localization, and miRNA target specificity especially 5' end variation [18]. Some studies revealed that isomiRs from a single precursor due to imprecise cleavage could be a way of broadening the regulatory network [21, 32, 35].

Here, in order to gain further insight into multiple isomiRs, we intended to study human isomiRs by reanalyzing public sequencing data using a strict analysis method.

The isomiR spectrum of miRNA was searched out based on known precursor sequence. Although many miRNAs are very well evolutionarily conserved phylogenetically across large evolutionary distances like vertebrates and fruit flies [11-13], sequence diversity of miRNA sequences across different animal species were identified by nucleotide variations and various

5' and/or 3' ends as well as species diversity [36,37]. Therefore, we also analyzed miRNA sequences in different animal species to discover potential relationship between miRNA sequence diversity and various isomiRs in *Homo sapiens*. Finally, because miRNAs in the miRBase database were detected by different methods, we also compared miRNAs across animal species based on high-throughput sequencing technologies.

Materials and Methods

Sequencing dataset of small RNAs in one human placental sample generated by SOLiD™ System were obtained from ABI (http://solidsoftwaretools.com/gf/project/srna/). We discarded any reads without a perfect 6 nt adaptor subsequence of 16 nt starting at the 5' end of read. The sequencing reads were aligned to all the miRNA precursor sequences of *Homo sapiens* in the miRBase database (Release 19.0, http://www.mirbase.org/) by tolerating one mismatch. Indeed, the only one mismatch was considered as a zero mismatch because SOLiD™ System employed ligation based chemistry with di-base labeled probes and each base was interrogated twice. Based on strict analysis, some potential miRNA* sequences could be detected according to known precursor sequences. Multiple isomiRs were found according to location of known miRNA and miRNA* sequence within the precursor. Here, any sequence mapped to precursor but observed less than five counts would be discarded from the study. All next generation sequencing technologies give sequences derived from a single molecule sampled from a larger population [38]. Therefore, sequence counts of isomiRs can reveal their relative expression levels. Moreover, miRNA sequences in different animal species were aligned with Clustal X 2.0 [39] by using the multiple alignment.

Results

Sequencing of Short RNAs and Analysis of isomiR Spectrum

After removing reads that had no perfect 6 nt adaptor sequences or had perfect 11 nt adaptor sequences in the region 1-25 nt of reads, a total of 7,080,817 reads remained These reads were mapped to human miRNA precursor sequences from the miRBase database to discover all kinds of isomiRs. After accurately mapping to precursor sequences, 1,948,608 (27.52%) reads were located on these precursors. A total of 2,633 kinds of short RNAs were found to locate on 313 kinds of miRNA precursor sequences. The phenomenon of multiple isomiRs could be easily detected based on miRNA precursor sequences (Figure 1). 264 precursors (84.35%) were found to have multiple isomiRs (2-60) with different sequence counts (5-176,285), lengths (16-27 nt), and 5' and/or 3' ends. 64.54 percent precursors yielded less than 14 kinds of isomiRs (Figure S1). But the expression level of miRNA was not directly related to the number of isomiR class. For example, the most abundant isomiR of hsa-miR-451 was sequenced 176,285 times, and 30 kinds of isomiRs were found. However, the most abundant isomiR of hsa-miR-1826 was sequenced 1,283 times, 60 kinds of isomiRs were identified. These star isomiRs always were involved more flexible 5' and/or 3' ends than isomiR class of mature miRNAs. The most abundant isomiR from a single precursor might be inconsistent with miRNA sequence in the miRBase database (Table 1). In the study, statistical analysis of 59 kinds of precursors showed that 38.98% precursors yielded miRNA sequences as the most abundant isomiRs. Other precursors yielded novel miRNA sequences as the most abundant isomiRs that might change 5' and/or 3' ends. Most these novel sequences were 3' isomiRs with the same seed sequences, and 7 kinds of them were 5' isomiRs that had novel seed sequences by "seed shifting" events (Table 1). Interestingly, we found 83.33% of these novel dominant sequences always were shorter than miRNA sequences. Compared with miRNA sequences in the miRBase database, they always lost 1-3 nucleotides at 5' and/or 3' ends especially at 3' end (Table 1). Although multiple isomiRs could be generated by changing 5' and/or 3' ends, terminus variation was easy to be found at 3' end (Figure 2). A single miRNA always was detected more 3' isomiRs than 5' isomiRs. 3' terminus variation was flexible and involved more nucleotides increased or decreased to produce novel 3' end (even the 3' ends were involved in additional non-template nucleotides). Generally, although 2-60 isomiRs were detected, a single precursor could yield several abundant isomiRs characterized by a narrow length range (20-23 nt) through dominant cleavage sites of Dicer and Drosha (Figure 3 and Figure 4). Some of them were found to have 3' additional non-template nucleosides, but these isomiRs with 3' additions always were not abundantly expressed in the miRNA locus. Almost all functional miRNAs could be found to have 1-3 abundant isomiRs that were 3' isomiRs with the same seed sequences. Some precursors, such as hsa-mir-199a-1 and hsa-mir-455, could yield 1-3 abundant isomiRs from different arms (Figure 3), because these special precursors could produce two kinds of functional miRNAs (miR-#-5p and miR-#-3p). Compared with these abundant isomiRs, other isomiRs always had fewer counts that were close to sequence counts of isomiRs of miRNA* (Figure 3).

Despite of isomiRs were produced by multiple cleavage sites of Dicer and Drosha, 1-3 continuous dominant cleavage sites always were identified by a narrow region (Figure 4).

Various miRNA Sequences in Different Animal Species

Various isomiRs from a single precursor were similar to various miRNA sequences in different animal species. In order to discover potential relationship between human isomiRs and miRNAs in other animal species, we analyzed some miRNA sequences across the animal kingdom. As shown in Figure 5, based on experimental data of miRNA sequences in different animal species from the miRBase database, these miRNAs were identified by nucleotide variations and various 5' and/or 3' ends (Figure 5). Some miRNA sequences (for example, hsa-miR-23a and hsa-miR-24) were common sequences that could be detected in many other animals, while other miRNA sequences were rare sequence, such as cfa-miR-23a, only was detected as private sequence in *Canis familiaris*. Interestingly, these miRNA sequences without nucleotide variation could be detected as human isomiRs with different sequence counts, and many of them were abundant isomiRs (Figure 5).

On the other hand, due to miRNAs in the miRBase were identified by different methods, especially the most abundant isomiR might be inconsistent with its reference sequence, we also compared sequence of the most abundant isomiR across different animal species through deep sequencing data based on published literatures. Some miRNAs had different sequences of the most abundant isomiRs across different animal species. These sequences were similar to multiple 5' isomiRs, and they always had the same seed sequences but various 3' ends.

Table 1. Relationship between the most abundant isomiR and human reference miRNA sequence from the miRBase database

Name	Start (abun/miR)	End (abun/miR)	Length (nt) (abun/miR)	Count (abun/miR)
hsa-mir-17	14	36	23	3147
hsa-mir-19b-1	54	76	23	3919
hsa-mir-24-1	44	65	22	160472
hsa-mir-26b	12	32	21	2077
hsa-mir-27b	61	81	21	1727
hsa-mir-29a	42	63	22	82625
hsa-mir-29c	54	75	22	45286
hsa-mir-30b	17	38	22	7067
hsa-mir-34a	22	43	22	1343
hsa-mir-103-1	48	70	23	8923
hsa-mir-125b-1	15	36	22	3140
hsa-mir-126	15	35	21	2968*
hsa-mir-130a	55	76	22	2201
hsa-mir-145	16	38	23	60143
hsa-mir-151	11	31	21	1635
hsa-mir-199a-1	6	28	23	10343
hsa-mir-218-1	25	45	21	3877
hsa-mir-337	23	43	21	1263
hsa-mir-376c	43	63	21	13152
hsa-mir-518c	62	84	23	3043
hsa-mir-521-1	54	75	22	6821
hsa-mir-525	15	35	21	1888
hsa-mir-526b	14	36	23	1091
hsa-mir-21	8	30/29	23/22	9199/3327
hsa-mir-23a	45	64/65	20/21	81817/65026
hsa-mir-23b	58	77/78	20/21	21365/9253
hsa-mir-27a	51	70/71	20/21	16310/12829
hsa-mir-29b-1	51	72/73	22/23	10843/6065
hsa-mir-30a	6	26/27	21/22	9279/1329
hsa-mir-30c-1	17	37/39	21/23	3647/1832

Table 1. (Continued)

Name	Start (abun/miR)	End (abun/miR)	Length (nt) (abun/miR)	Count (abun/miR)
hsa-mir-30d	6	26/27	21/22	8160/32
hsa-mir-99a	13	33/34	21/22	5456/68
hsa-mir-100	13	33/34	21/22	31862/86
hsa-mir-125a	15	36/38	22/24	3804/22
hsa-mir-143	61	80/81	20/21	13466/4750
hsa-mir-181a-1	24	44/46	21/23	1046/879
hsa-mir-191	16	35/38	20/23	8942/754
hsa-mir-205	34	54/55	21/22	1250/297
hsa-mir-223	68	90/89	23/22	2066/311
hsa-mir-376b	62	82/83	21/22	1609/35
hsa-mir-424	11	31/32	21/22	25663/15
hsa-mir-451	17	37/38	21/22	176285/107762
hsa-mir-515-1	51	71/72	21/22	8503/5634
hsa-mir-516b-1	16	36/37	21/22	1266/427
hsa-mir-518a-1	14	34/33	21/20	1448/100
hsa-mir-519d	54	76/75	23/22	4380/637
hsa-mir-520d	16	36/35	21/20	4992/212
hsa-mir-520g	55	75/78	21/24	1177/8
hsa-mir-524	16	36/37	21/22	2875/44
hsa-mir-542	53	73/74	21/22	1411/133
hsa-mir-574	61	81/82	21/22	1735/604
hsa-mir-1826	1	21/27	21/27	1283/—
hsa-mir-136	50/49	70	21/22	2872*/2379*
hsa-mir-517a	55/54	75	21/22	66880/4682
hsa-mir-517c	58/57	78	21/22	10146/930
hsa-mir-519a-1	54/53	74	21/22	1782/434
hsa-mir-140	63/62	84/82	22/21	2513/45
hsa-mir-487b	53/51	73/72	21/22	7759/1891
hsa-mir-512-1	15/14	35/36	21/23	12303/299

The above listed most abundant isomiR are sequenced more than 1000 times (sequence counts >1000). It includes the precursor, location (start and end positions on precursor) of the most abundant isomiR and miRNA sequence in the miRBase database, and their sequence counts.

abun/miR: the most abundant isomiR/miRNA sequence. * the sequence is identifed as miRNA* in the miRBase database. — The miRNA sequence has not detected or less than 5 sequence counts.

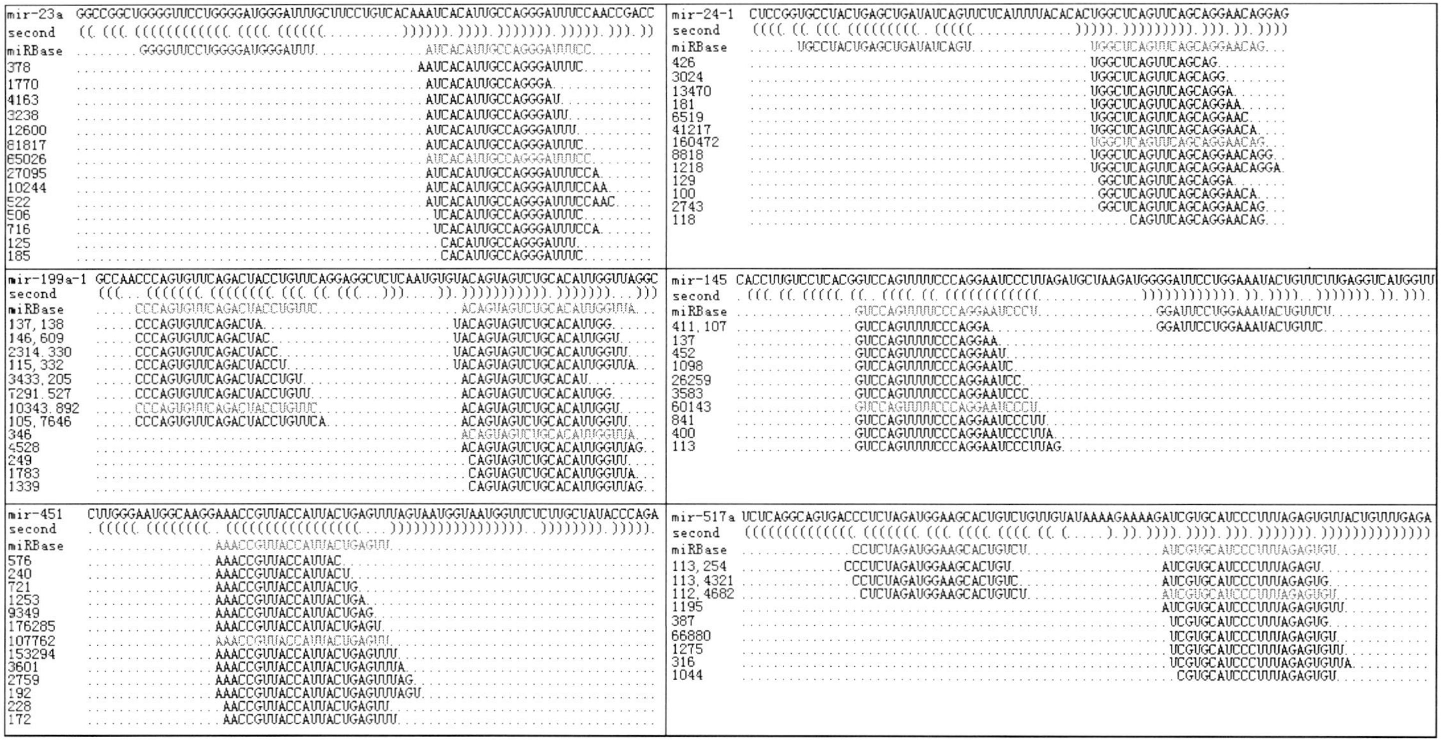

Figure 1. The repertoire of isomiRs based on miRNA precursors. The precursor sequence, second structure of precursor (stem-loop), mature (red) and its miRNA* sequence (blue) in the miRBase database, and sequence counts of different isomiRs are shown in the left column. Hsa-mir-199a-1 yields two kinds of functional miRNA sequences: hsa-miR-199a-5p and hsa-miR-199a-3p. The isomiRs with counts less than 100 are not included.

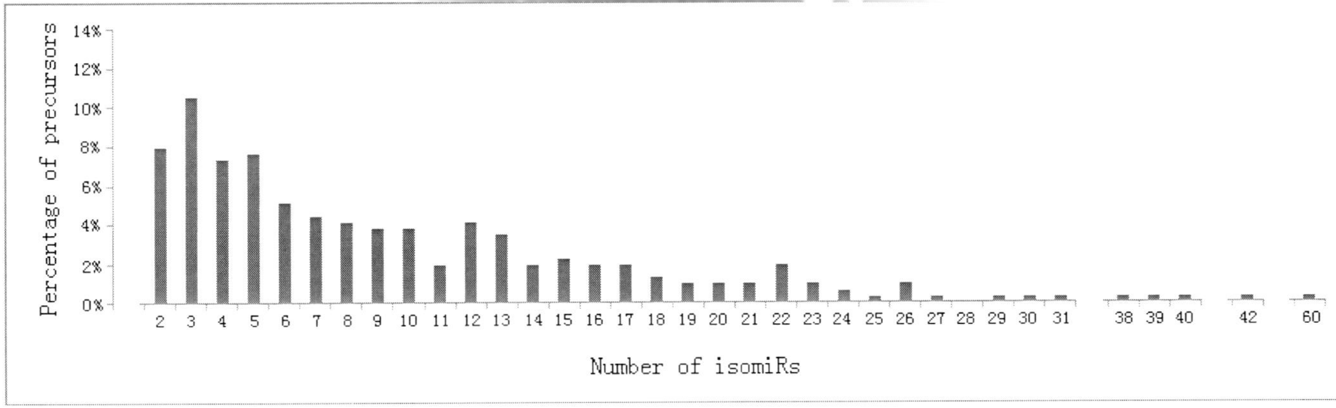

Figure S1. Distribution of isomiR number in the whole precursors in human tissue. Here we do not include the 49 precursors which have only one kind of miRNA sequence or have two kinds of sequences that are mature miRNA and miRNA* sequences or miR-#-5p and miR-#-3p. The number of isomiRs shows all isomiRs of miRNA and miRNA*.

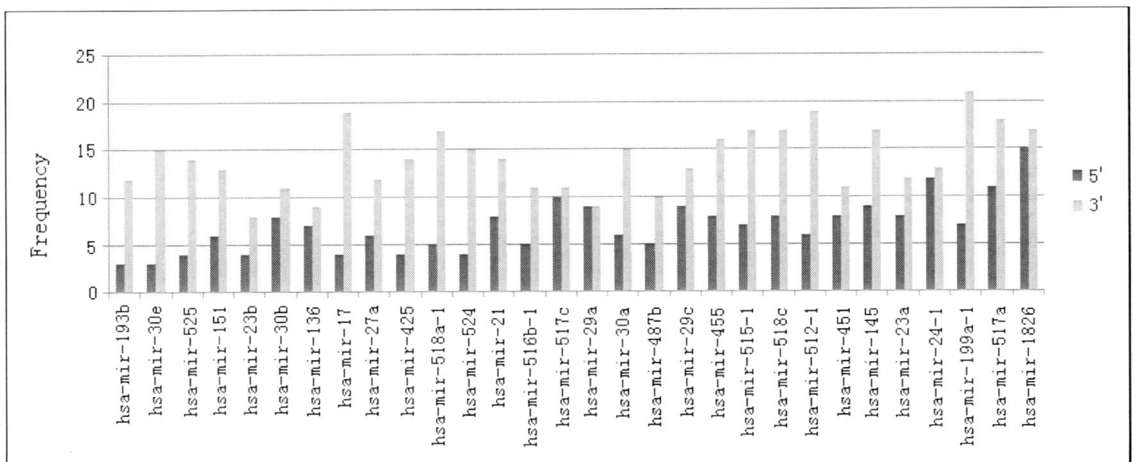

Figure 2. Frequencies of variation of isomiRs with two different ends in 30 miRNA precursors. Each precursor yields different numbers of isomiRs with different the 5' and/or 3' ends. The precursors which produce less than 20 kinds of isomiRs are not included.

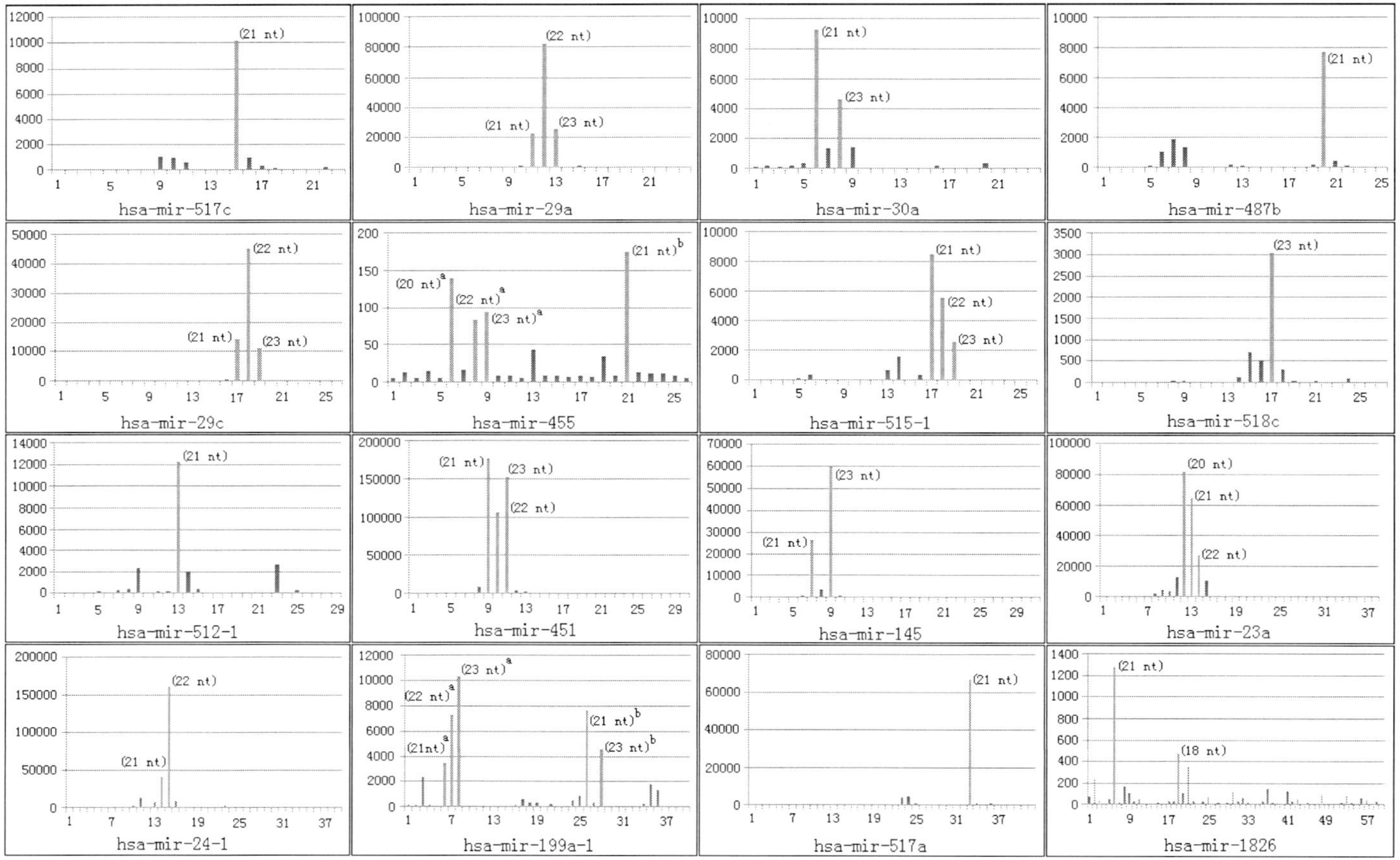

Figure 3. Comparison of relative expression levels of isomiRs in 16 miRNA precursors. Vertical axis: the sequence counts; horizontal axis: number of isomiRs. The lengths of abundant isomiRs (yellow columns) are presented. [a] these abundant isomiRs with the same seed sequences are produced by 5' arm (5p) of precursor (such as hsa-mir-455 and hsa-mir-199a-1); [b] the abundant isomiRs with the same seed sequences are produced by 3' arm (3p) of the precursor. Abundant isomiRs of other precursors are characterized as 3' isomiRs which had the same seed sequences.

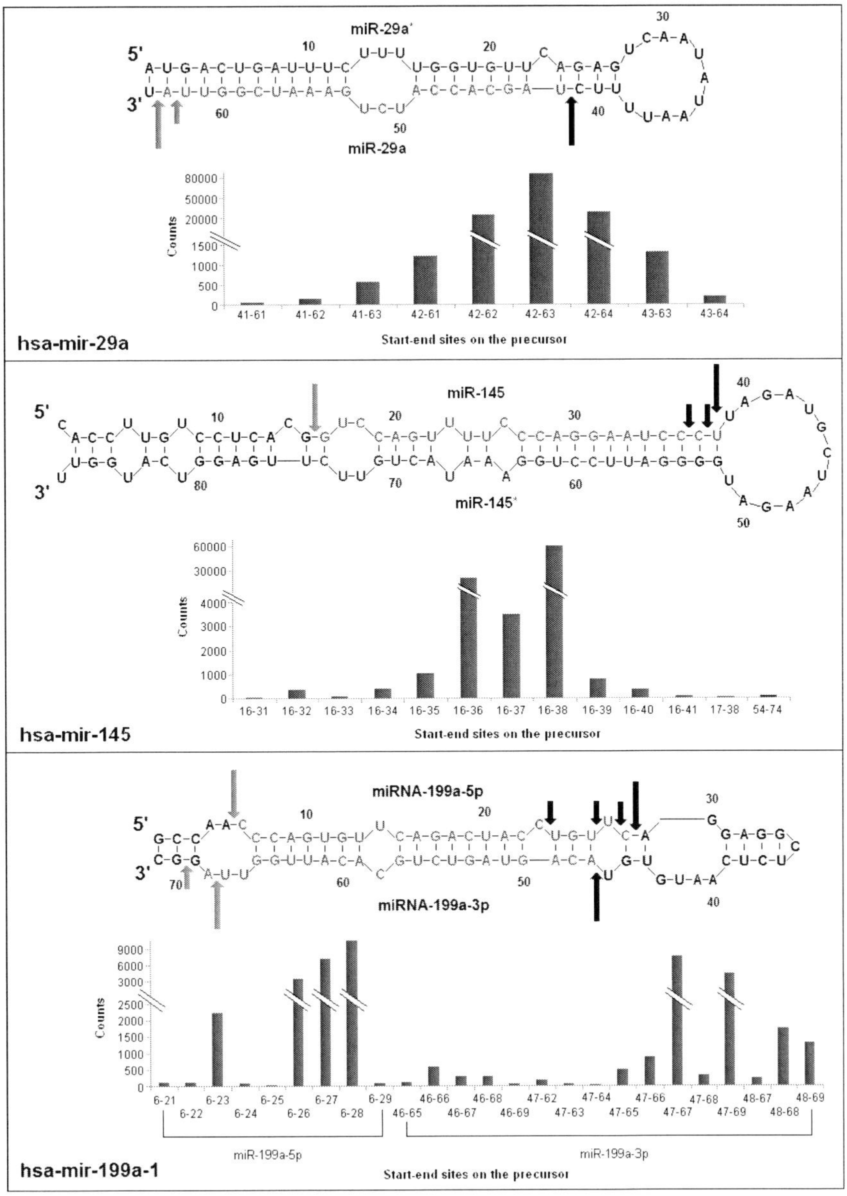

Figure 4. Dominant cleavage sites of Drosha and Dicer and relative expression level of isomiRs. Mature miRNA and its isomiRs are shown in red, and miRNA* and its isomiRs are shown in blue. The mature miRNA and its star sequence in stem-loop are miRNA sequences from the miRBase database. Inferred cleavage sites of Drosha and Dicer are indicated in grey and dark arrows, respectively. Large arrows represent the dominant cleavage sites for the most abundant isomiR whereas small arrows indicate other abundant isomiRs. Relative expression levels of all kinds of isomiRs (counts > 50) are shown based on their sequence counts. Both hsa-miR-199a-5p and hsa-miR-199a-3p are reported as functional mature miRNAs. All of these isomiRs are produced by different cleavage sites of Drosha and Dicer on the precursor, and their start-end sites are presented corresponding cleavage sites.

Name	Sequence	Count	Name	Sequence	Count	Name	Sequence	Count
bta-miR-21	UAGCUUAUCA GACUGAUGUU GACU	8	bta-miR-23a	AUCACAUUGC CAGGGAUUUC CA	27095	cfa-miR-24	UGGCUCAGUU CAGCAGGAAC AGG	8818
cfa-miR-21	....--	3327	dre-miR-23a	....		bta-miR-24-3p	....-	160472
cgr-miR-21	....--		oan-miR-23a	....		dre-miR-24	....-	
gga-miR-21	....--		hsa-miR-23a	....-	65026	gga-miR-24	....-	
hsa-miR-21	....--		mmu-miR-23a	....-		hsa-miR-24	....-	
mmu-miR-21	....--		rno-miR-23a	....-		mmu-miR-24	....-	
oan-miR-21	....--		ssc-miR-23a	....-		oan-miR-24	....-	
rno-miR-21	....--		mdo-miR-23a	.... --	81817	rno-miR-24	....-	
ssc-miR-21	....--		cfa-miR-23a	....- --	12600	ssc-miR-24	....-	
dre-miR-21	....G.... .G.-					mdo-miR-24	....---- ---	426

Name	Sequence	Count	Name	Sequence	Count	Name	Sequence	Count
bta-miR-29a	CUAGCACCAU CUGAAAUCGG UUA-	603	bta-miR-29c	UAGCACCAUU UGAAAUCGGU UA	80175	bta-miR-30a-5p	UGUAAACAUC CUCGACUGGA AGCU	1477
cfa-miR-29a	-......... ....-	82625	cfa-miR-29c	....		cfa-miR-30a	....-	4649
hsa-miR-29a	-......... ....-		hsa-miR-29c	....		oan-miR-30a	....-	
mmu-miR-29a	-......... ....-		mmu-miR-29c	....		gga-miR-30a-5p	....--	1329
rno-miR-29a	-......... ....-		rno-miR-29c	....		hsa-miR-30a	....--	
dre-miR-29a	-......... U......... ...-		gga-miR-29c	.... --	491	mmu-miR-30a	....--	
cap-miR-29a	-......... U......A... ..U-					rno-miR-30a	....--	
sko-miR-29a	-......... A......A... ..U-					ssc-miR-30a	....--	
oan-miR-29a	-......... U......... ..--					dre-miR-30a	.........U .C........ ..--	
bfl-miR-29a	-......... A......A... ...U							

Name	Sequence	Count	Name	Sequence	Count	Name	Sequence	Count
mml-miR-145	GUCCAGUUUU CCCAGGAAUC CCUU	841	hsa-miR-199a-5p	CCCAGUGUUC AGACUACCUG UUC	10343	bta-miR-451	AAACCGUUAC CAUUACUGAG UUU	153294
ssc-miR-145	....		mmu-miR-199a-5p	....		gga-miR-451	....	
bta-miR-145	....-	60143	rno-miR-199a-5p	....		oan-miR-451	....	
hsa-miR-145	....-		xtr-miR-199a	....		dre-miR-451	....-	107762
mdo-miR-145	....-		bta-miR-199a-5p	....-	7291	hsa-miR-451	....-	
mmu-miR-145	....-		bta-miR-199a-3p	-A....AG.. U.CAC.UUG. ..A	346	mmu-miR-451	....-	
rno-miR-145	....-		hsa-miR-199a-3p	-A....AG.. U.CAC.UUG. ..A		rno-miR-451	....-	
dre-miR-145	....--	3583	mmu-miR-199a-3p	-A....AG.. U.CAC.UUG. ..A				
oan-miR-145	....- ----	452	rno-miR-199a-3p	-A....AG.. U.CAC.UUG. ..A				

Figure 5. Sequences of miRNAs across animal species and their counts as human isomiRs. All of these miRNA sequences of different animal species are experimental data from the miRBase database. Most these sequences can be also found as human isomiRs.
The counts of these human isomiRs are given and labeled at the first sequence. Both miR-199a-5p and miR-199a-3p are reported as functional mature miRNAs. Sequences identical to the first sequence are indicated with dots, and missing sites are indicated with dashes (-).

Discussion

By reanalyzing multiple human isomiRs based on strict analysis from sequencing data using SOLiD^{TM} system, we found 84.35 percent miRNA precursors were detected to have 2-60 isomiRs with different sequence counts, which revealed imprecise cleavage of Dicer and Drosha was a widespread phenomenon. Indeed, the various sequences with 5' and/or 3' modification had been reported in many animals and plants [18-28,40-42], such as *Caenorhabditis elegans* [24], *Mus musculus* [43], *Bombyx mori* [26], *Pinus contorta* and *Oryza sativa* [44], and etc. Most precursors (64.54%) yielded less than 14 kinds of isomiRs, while some precursors expressed at high levels could yield more isomiRs based on moderate correlation [20]. Despite of multiple isomiRs from a single precursor, less miRNA* were found to have isomiRs. According to miRNA biogenesis, a hairpin precursor processing by Dicer can yield three products: the loop of the hairpin, functional mature miRNA and miRNA* [45]. The loop sequence is degraded as a by-product, and star sequence always is typically degraded like the loop, whereas the mature miRNA strand is taken up into the microribonucleoprotein complex (miRNP) [46]. It is plausible that many star isomiRs might be degraded like their miRNA* and not sequenced. Nonetheless, we still found some star isomiRs, for example, hsa-mir-517a yielded 11 kinds of star isomiRs with fewer counts and flexible 5' and/or 3' ends. Although most of precursors yielded 2-60 isomiRs with different sequence counts, another plausible factor should not be ignored. These precursors can also produce various short RNAs like various isomiRs either through non-Dicer processing or degradation. Due to only fragments of at least 16 nt were involved in the study, a great amount of shorter metabolic products were not detected. When these by-products are mapped back to the secondary structure, these various short RNAs termed as special "isomiRs" might contribute to multiple isomiRs though they might have fewer sequence counts. Interestingly, except several kinds of abundant isomiRs from a single precursor, others always had fewer sequence counts like star isomiRs (Figure 3). Consequently, we believed that the phenomenon of multiple isomiRs might not only result from the imprecise and alternative cleavage, but also from special "isomiRs" produced by hairpins through non-Dicer processing or degradation. These abundantly expressed isomiRs might play important roles in biological process (discussed later) although the most abundant isomiR may be a better marker to analyze relative expression levels of miRNAs [47], whereas other rare isomiRs might be by-products during imprecise pre-miRNA processing or non-Dicer processing or degradation of hairpins.

miRNA sequence was modified at 5' and/or 3' ends to be multiple isomiRs, and the seed sequences of 5' isomiRs were shifted by "seed shifting" events. It is found that 3' ends variation was much more common than 5' ends (Figure 2) [20,21,32], and fewer precursors selected 5' isomiRs as functional miRNA sequences, which might imply cleavage bias of Dicer during pre-miRNA processing. 3' isomiRs changed their 3' ends but still remained the same seed sequences to bind target mRNAs, while 5' isomiRs would change the identity of the seed sequences by shifting toward 5' or 3' through "seed shifting" events. Therefore, imprecise cleavage at 5' end would endure through strict regulation or selection of functional pressure. Indeed, almost all abundant isomiRs were 3' isomiRs and had the same seed sequences. All of these results could be explained with cleavage bias of Dicer during pre-miRNA processing. The cleavage bias might be an optimal choice that resulted from functional selection pressure that is a driving force in avoiding generating novel 5' isomiRs with new seed sequences. The process would be involved in complex regulatory networks to interfere with pre-miRNA processing, or there were special structure or sites or flanking sequence of precursor to restrict Dicer and Drosha activity. In addition, almost all miRNAs were found to have 1-3 abundant isomiRs characterized by a narrow range of length (20-23 nt) (Figure 3). The length distribution of abundant isomiRs revealed that length of 20-23 nt had a selective advantage over others and verified advantage cleavage based on cleavage bias of Dicer and Drosha (Figure 4). The length of abundant isomiRs confirmed that ~22 nt miRNA was the most appropriate length to bind target mRNAs. According to length diversity of the most abundant isomiRs, it is plausible that the lengths of miRNAs could influence directly or indirectly regulating biological process. Generally, in a single precursor, the difference of lengths of abundant isomiRs was 1-2 nt and advantage cleavage sites might be a flexible region of 2-3 continuous sites with cleavage bias at 3' ends (Figure 4). As the final products, several kinds of abundant isomiRs particularly fewer 5' isomiRs might play important roles in biological process as functional miRNAs, and could be a way of broadening the network of factors and processes that are regulated by miRNAs [21,32,35].

In these abundant isomiRs, expression level of the most abundant isomiR always was quite remarkable (Figure 3), and it should be easily cloned by experimental approaches as functional mature miRNA sequence. All of these results revealed that the phenomenon of multiple isomiRs were non-random event.

In order to get more potential information of isomiRs, combining similarity between multiple isomiRs from a single precursor and sequence diversity of corresponding miRNAs across animal species, we also analyzed some miRNAs in different animal species (Figure 5). Based on these miRNA sequences, the miRNA sequences may be various in different animals by changing 5' and/or 3' ends and nucleotide variations as species diversity (Figure 5) [36,37]. The sequence diversity might result from different selective advantages. Intriguingly, all of these different miRNA sequences without nucleotide variation were detected as human isomiRs with different sequence counts, and many of them were abundant isomiRs (Figure 5). Generally, miRNAs in the miRBase database should be the most abundant isomiRs that can be easily detected despite they may be detected by different experimental methods based on different experimental samples. Simultaneously, in order to avoid influence of different experimental methods, according to published sequencing data, we also analyzed the most abundant isomiR spectrum across different animal species.

Expectably, we found different animal species maybe have different sequences of the most abundant isomiRs. Combining the universal imprecise cleavage during pre-miRNA processing, we speculated that different animals might select different sequences as the most abundant isomiRs that could be detected by experimental approaches. It is plausible that the cellular environment and evolutionary and functional implication might contribute to selective pressure and might be a critical decision during producing the relative abundant functional miRNA sequence.

Taken together, a detailed analysis of the isomiRs from small RNA sequencing dataset of *Homo sapiens* was carried out. Our studies suggested that most miRNA loci yielded multiple isomiRs with different sequence counts and lengths. Although imprecise cleavage of Dicer and Drosha was a universal phenomenon, hairpins through non-Dicer processing or degradation also contributed to the phenomenon of multiple isomiRs, particularly those rare isomiRs with fewer sequence counts. 3' terminus variation was much more common than 5' based on cleavage bias of Dicer and Drosha. Most isomiRs were 3' isomiRs with the same seed sequences, while the 5' isomiRs were found to have "seed shifting" phenomenon, especially shifting toward 3' terminus. Most miRNAs had a set of isomiRs, in which several were abundant isomiRs.

These abundant isomiRs, characterized by a narrow length distribution (20-23 nt), might act as functional miRNAs and play important roles in biological processes. Experimental miRNA sequences in other animal species could be detected as human abundant isomiRs. These findings revealed that the most abundant isomiR spectrum might change in different animal species, even in different tissues and/or at different times due to change of cellular environment. In order to regulate biological process more flexibly and effectively, organism always produces several abundant isomiRs from a given miRNA locus, which might be an optimal choice to select dominant miRNA sequence to adapt to complex cellular environment and evolutionary and functional selective pressures.

Acknowledgments

The work was supported by National Natural Science Foundation of China (No. 81072389), the Research Found for the Doctoral Program of Higher Education of China (No. 211323411002), China Postdoctoral Science Foundation funded project (No. 2012M521100), key Grant of Natural Science Foundation of the Jiangsu Higher Education Institutions of China (No. 10KJA33034), Jiangsu Planned Projects for Postdoctoral Research Funds (No. 1201022B), Science and Technology Development Fund Key Project of Nanjing Medical University (No. 2012NJMU001), and the Priority Academic Program Development of Jiangsu Higher Education Institutions (PAPD).

References

[1] Bartel DP, Chen CZ (2004) Micromanagers of gene expression: the potentially widespread influence of metazoan microRNAs. *Nature Reviews Genetics* 5: 396-400.
[2] Plasterk RHA (2006) Micro RNAs in animal development. *Cell* 124: 877-881.
[3] Cai XZ, Hagedorn CH, Cullen BR (2004) Human microRNAs are processed from capped, polyadenylated transcripts that can also function as mRNAs. *Rna-a Publication of the Rna Society* 10: 1957-1966.
[4] Lund E, Guttinger S, Calado A, Dahlberg JE, Kutay U (2004) Nuclear export of microRNA precursors. *Science* 303: 95-98.
[5] O'Toole AS, Miller S, Haines N, Zink MC, Serra MJ (2006) Comprehensive thermodynamic analysis of 3 ' double-nucleotide overhangs neighboring Watson-Crick terminal base pairs. *Nucleic Acids Research* 34: 3338-3344.
[6] Okamura K, Phillips MD, Tyler DM, Duan H, Chou YT, et al. (2008) The regulatory activity of microRNA star species has substantial influence on microRNA and 3 ' UTR evolution. *Nature Structural and Molecular Biology* 15: 354-363.
[7] Ro S, Park C, Young D, Sanders KM, Yan W (2007) Tissue-dependent paired expression of miRNAs. *Nucleic Acids Research* 35: 5944-5953.

[8] Jagadeeswaran G, Zheng Y, Sumathipala N, Jiang HB, Arrese EL, et al. (2010) Deep sequencing of small RNA libraries reveals dynamic regulation of conserved and novel microRNAs and microRNA-stars during silkworm development. *Bmc Genomics* 11: 52.

[9] Guo L, Lu Z (2010) The Fate of miRNA* Strand through Evolutionary Analysis: Implication for Degradation as Merely Carrier Strand or Potential Regulatory Molecule? *PloS ONE* 5: e11387.

[10] Siomi H, Siomi MC (2009) On the road to reading the RNA-interference code. *Nature* 457: 396-404.

[11] Lagos-Quintana M, Rauhut R, Lendeckel W, Tuschl T (2001) Identification of novel genes coding for small expressed RNAs. *Science* 294: 853-858.

[12] *Lau NC*, Lim LP, Weinstein EG, Bartel DP (2001) An abundant class of tiny RNAs with probable regulatory roles in Caenorhabditis elegans. *Science* 294: 858-862.

[13] Lee RC, Ambros V (2001) An extensive class of small RNAs in Caenorhabditis elegans. *Science* 294: 862-864.

[14] Bartel DP (2004) MicroRNAs: Genomics, biogenesis, mechanism, and function. *Cell* 116: 281-297.

[15] Grimson A, Srivastava M, Fahey B, Woodcroft BJ, Chiang HR, et al. (2008) Early origins and evolution of microRNAs and Piwi-interacting RNAs in animals. *Nature* 455: 1193-1197.

[16] Liu N, Okamura K, Tyler DM, Phillips MD, Chung WJ, et al. (2008). The evolution and functional diversification of animal microRNA genes. *Cell Research* 18: 985-996.

[17] Niwa R, Slack FJ (2007) The evolution of animal microRNA function. *Current Opinion in Genetics and Development* 17: 145-150.

[18] Borel C, Antonarakis SE (2008) Functional genetic variation of human miRNAs and phenotypic consequences. *Mammalian Genome* 19: 503-509.

[19] Cummins JM, He YP, Leary RJ, Pagliarini R, Diaz LA, et al. (2006) The colorectal microRNAome. *Proceedings of the National Academy of Sciences of the United States of America* 103: 3687-3692.

[20] Kuchenbauer F, Morin RD, Argiropoulos B, Petriv OI, Griffith M, et al. (2008) In-depth characterization of the microRNA transcriptome in a leukemia progression model. *Genome Research* 18: 1787-1797.

[21] Landgraf P, Rusu M, Sheridan R, Sewer A, Iovino N, et al. (2007) A mammalian microRNA expression atlas based on small RNA library sequencing. *Cell* 129: 1401-1414.

[22] Morin RD, O'Connor MD, Griffith M, Kuchenbauer F, Delaney A, et al. (2008) Application of massively parallel sequencing to microRNA profiling and discovery in human embryonic stem cells. *Genome Res* 18: 610-621.

[23] Rathjen T, Pais H, Sweetman D, Moulton V, Munsterberg A, et al. (2009) High throughput sequencing of microRNAs in chicken somites. *Febs Letters* 583: 1422-1426.

[24] Ruby JG, Jan C, Player C, Axtell MJ, Lee W, et al. (2006) Large-scale sequencing reveals 21U-RNAs and additional microRNAs and endogenous siRNAs in C-elegans. *Cell* 127: 1193-1207.

[25] Wu H, Neilson J, Kumar P, Manocha M, Shankar P, et al. (2007) miRNA profiling of naive, effector and memory CD8 T cells. *PLoS ONE* 2: e1020.

[26] Yu X, Zhou Q, Li S-C, Luo Q, Cai Y, et al. (2008) The Silkworm (Bombyx mori) microRNAs and Their Expressions in Multiple Developmental Stages. *PLoS ONE* 3: e2997.

[27] Ahn HW, Morin RD, Zhao H, Harris RA, Coarfa C, et al. (2010) MicroRNA transcriptome in the newborn mouse ovaries determined by massive parallel sequencing. *Molecular Human Reproduction* 16: 463-471.

[28] Liao JY, Ma LM, Guo YH, Zhang YC, Zhou H, et al. (2010) Deep Sequencing of Human Nuclear and Cytoplasmic Small RNAs Reveals an Unexpectedly Complex Subcellular Distribution of miRNAs and tRNA 3 ' Trailers. *Plos One* 5: e10563.

[29] Guo L, Yang Q, Lu JF, Li HL, Ge QY, et al. (2011) A Comprehensive Survey of miRNA Repertoire and 3 ' Addition Events in the Placentas of Patients with Pre-Eclampsia from High-Throughput Sequencing. *Plos One* 6: e21072.

[30] Li SC, Liao YL, Chan WC, Ho MR, Tsai KW, et al. (2011) Interrogation of rabbit miRNAs and their isomiRs. *Genomics* 98: 453-459.

[31] Guo L, Li H, Liang T, Lu J, Yang Q, et al. (2012) Consistent isomiR expression patterns and 3' addition events in miRNA gene clusters and families implicate functional and evolutionary relationships. *Mol. Biol. Rep.* 39: 6699-6706.

[32] Morin RD, O'Connor MD, Griffith M, Kuchenbauer F, Delaney A, et al. (2008) Application of massively parallel sequencing to microRNA profiling and discovery in human embryonic stem cells. *Genome Research* 18: 610-621.

[33] Guo L, Yang Q, Lu J, Li H, Ge Q, et al. (2011) A Comprehensive Survey of miRNA Repertoire and 3' Addition Events in the Placentas of Patients with Pre-eclampsia from High-throughput Sequencing. *PLoS ONE* 6: e21072.

[34] Guo L, Li H, Lu J, Yang Q, Ge Q, et al. (2011) Tracking miRNA precursor metabolic products and processing sites through completely analyzing high-throughput sequencing data. *Molecular Biology Reports* 39: 2031-2038.

[35] Azuma-Mukai A, Oguri H, Mituyama T, Qian ZR, Asai K, et al. (2008) Characterization of endogenous human Argonautes and their miRNA partners in RNA silencing. *Proceedings of the National Academy of Sciences of the United States of* America 105: 7964-7969.

[36] Guo L, Sun BL, Sang F, Wang W, Lu ZH (2009) Haplotype Distribution and Evolutionary Pattern of miR-17 and miR-124 Families Based on Population Analysis. *PLoS ONE* 4: e7944.

[37] Wheeler BM, Heimberg AM, Moy VN, Sperling EA, Holstein TW, et al. (2009) The deep evolution of metazoan microRNAs. *Evolution and Development* 11: 50-68.

[38] Alg HK, Li K, Chen ZP, Pang JCS (2008) MicroRNA-124 functions as a potential tumor suppressor in medulloblastoma. *Modern Pathology* 21: 322a.

[39] Larkin MA, Blackshields G, Brown NP, Chenna R, McGettigan PA, et al. (2007) Clustal W and clustal X version 2.0. *Bioinformatics* 23: 2947-2948.

[40] Fernandez-Valverde SL, Taft RJ, Mattick JS (2010) Dynamic isomiR regulation in Drosophila development. *RNA* 16: 1881-1888.

[41] Burroughs AM, Ando Y, de Hoon MJ, Tomaru Y, Nishibu T, et al. (2010) A comprehensive survey of 3' animal miRNA modification events and a possible role for 3' adenylation in modulating miRNA targeting effectiveness. *Genome Res* 20: 1398-1410.

[42] Lee LW, Zhang S, Etheridge A, Ma L, Martin D, et al. (2010) Complexity of the microRNA repertoire revealed by next generation sequencing. *RNA* 16: 2170-2180.

[43] Reid JG, Nagaraja AK, Lynn FC, Drabek RB, Muzny DM, et al. (2008) Mouse let-7 miRNA populations exhibit RNA editing that is constrained in the 5'-seed/cleavage/anchor regions and stabilize predicted mmu-let-7a: mRNA duplexes. *Genome Research* 18: 1571-1581.

[44] Morin RD, Aksay G, Dolgosheina E, Ebhardt HA, Magrini V, et al. (2008) Comparative analysis of the small RNA transcriptomes of Pinus contorta and Oryza sativa. *Genome Research* 18: 571-584.

[45] Friedlander MR, Chen W, Adamidi C, Maaskola J, Einspanier R, et al. (2008) Discovering microRNAs from deep sequencing data using miRDeep. *Nature Biotechnology* 26: 407-415.

[46] Filipowicz W, Bhattacharyya SN, Sonenberg N (2008) Mechanisms of post-transcriptional regulation by microRNAs: are the answers in sight? *Nature Reviews Genetics* 9: 102-114.

[47] Guo L, Lu Z (2010) Global expression analysis of miRNA gene cluster and family based on isomiRs from deep sequencing data. Computational *Biology and Chemistry* 34: 165-171.

Index

B

C

D

E

F

G

P

T

U

V

W

X

Y

Z